The Energy Transition

Vincent Petit

The Energy Transition

An Overview of the True Challenge of the 21st Century

 Springer

Vincent Petit
Grenoble, France

ISBN 978-3-319-50291-5   ISBN 978-3-319-50292-2   (eBook)
DOI 10.1007/978-3-319-50292-2

Library of Congress Control Number: 2016963066

Printed on acid-free paper

This Springer imprint is published by Springer Nature
The registered company is Springer International Publishing AG
The registered company address is: Gewerbestrasse 11, 6330 Cham, Switzerland

Foreword

In recent years, the world has gone through enormous changes. Enabled by better health services, the world's population has exploded. From 2.5 billion people in 1950, we now share the planet with over seven billion people, and it is predicted that this number will grow to nine billion by the middle of the century.

Along with the population explosion has come complete globalization and interconnectivity. Cross-continental travel, global trade, business specialization, and technology have gotten us, billions of people, to work together on a daily basis, to share our innovations, and to mutually benefit from one another. Living standards have soared in many regions, propelled by advances in innovation, health services, and mobility. Populations once isolated now embrace technologies they could not previously access, dramatically speeding up their industrial, technological, and economic evolution. As they connect to the global economy, they also contribute to its overall growth with their energy, their capabilities, and most importantly, their hope for a better life. In doing so, they elevate the overall living standards of the whole world. As we all live in this densely populated world, we've also come to the realization that we're all in this together. We share the same planet, the same resources, and the same climate. Whatever our differences, we are bound by a common responsibility to our common planet.

The foundation of all progress, indeed, of all life, is energy. Energy makes everything work: heating, lighting, water, education and learning, health, technology, manufacturing, and mobility. Energy consumption has increased by around 45 % in the last 20 years. And yet, this is only the beginning, as there are still many of us on this planet who do not have reliable access to energy, including more than one billion people who don't have any access to electricity at all. Energy consumption is set to increase by another 35 % within the next 20 years, a continuation of the development which started a few decades ago.

But this cannot go on the way it used to. Energy usage and generation are also the biggest reasons for climate change by being the largest sources of carbon emissions. Fossil fuels still represent over 80% of the total primary energy consumed in the world. With fossil fuels come greenhouse gas emissions, which have already reached an unprecedented level in the history of our planet and are still increasing. The consequences on sustainability are already visible, and will continue to worsen in the coming decades, if we do not act. Scientists have estimated that the level of emissions needs to be cut by 35% in the next 20 years to prevent catastrophic consequences.

This, combined with the forecasted energy increase of 35% in the same period, supposes that we invent ways to do what we do today at a carbon efficiency that would be significantly improved (over time up to three times better than today). Humanity is now facing a complicated equation to solve, which is starting to look like an energy deadlock. As we have now entered an era of abundance and economic growth, we require a permanent answer to how we can get more energy. At the same time, this surge in the demand is creating the conditions for massive climate disruption, which will impede or even halt humanity's progress.

Many of the major players affecting this change already understand the dramatic consequences of inaction and have put together elaborate plans and, in some instances, made commitments to tame their energy consumption or produce it in a cleaner manner. But the magnitude of the change requires the mobilization of all.

The purpose of the "Energy Transition" is to bring a simple and holistic view on this complex issue. If the problem is, at its very essence, a global one, the situations of each country and each industry will all be greatly different. The "Energy Transition" is an attempt to provide both a global and a local perspective to this proposed evolution.

But it's not just an overview. It also lists solutions to the problem. In each industry, in each country, the potential exists for a more efficient use and a cleaner production of energy. Looking even further, advancing technology and innovation now bring new perspectives which could radically change the energy paradigm in which we have been living with for decades. The impossible energy equation that the world is facing now has a chance to be cracked, and this constitutes as one of the biggest technology revolutions that our next generation needs to step up and lead, one that will create jobs and innovation opportunities. The "Energy Transition" raises both concern over the problem and confidence that this challenge can and will be met.

Our generation started the twenty-first century. It is our responsibility to build a future which is sustainable while enabling the development of all on our planet. The energy transition is the biggest challenge we face right now in this century. Resolved by technology and innovation, it can become another example of how human beings transform deadlocks into major opportunities, reinventing the world we live in.

Schneider Electric, Chairman & CEO   Jean-Pascal Tricoire

Contents

List of Countries

Asia (Non OECD)	Samoa, Bangladesh, Brunei Darussalam, Bhutan, Channel Islands, Fiji, Guam, Micronesia, Fed. Sts., Indonesia, Cambodia, Kiribati, Korea, Rep., Lao PDR, Sri Lanka, Maldives, Marshall Islands, Myanmar, Mongolia, Northern Mariana Islands, Malaysia, Nepal, Pakistan, Philippines, Palau, Papua, New Guinea, Singapore, Solomon Islands, Suriname, Timor-Leste, Thailand, Tuvalu, Taiwan, China, Vietnam, Vanuatu, Samoa
Africa	Aruba, Angola, Burundi, Benin, Burkina Faso, Botswana, Central African Republic, Cote d'Ivoire, Cameroon, Congo, Rep., Comoros, Cabo Verde, Djibouti, Algeria, Egypt, Arab Rep., Eritrea, Ethiopia, Gabon, Ghana, Guinea, Gambia, Guinea-Bissau, Equatorial Guinea, Kenya, Liberia, Libya, Lesotho, Morocco, Madagascar, Mali, Mozambique, Mauritania, Mauritius, Malawi, Namibia, Niger, Nigeria, Rwanda, Sao Tome and Principe, Sudan, Senegal, Sierra Leone, Somalia, South Sudan, Swaziland, Chad, Togo, Tonga, Tunisia, Tanzania, Uganda, South Africa, Congo, Dem. Rep., Zambia, Zimbabwe
Eurasia	Afghanistan, Armenia, Azerbaijan, Belarus, Georgia, Kazakhstan, Kyrgyz Republic, Russian Federation, Tajikistan, Turkmenistan, Uzbekistan
Europe	Andorra, Albania, Austria, Belgium, Bulgaria, Bosnia and Herzegovina, Central Europe and the Baltics, Switzerland, Cyprus, Czech Republic, Germany, Denmark, Spain, Estonia, Finland, France, Faeroe Islands, United Kingdom, Greece, Grenada, Greenland, Croatia, Hungary, Isle of Man, Ireland, Iceland, Italy, Kosovo, Liechtenstein, Lithuania, Luxembourg, Latvia, Monaco, Moldova, Macedonia, FYR, Malta, Montenegro, Netherlands, Norway, Poland, Portugal, Romania, Serbia, Slovak Republic, Slovenia, Sweden, Ukraine

Middle East	United Arab Emirates, Bahrain, Iran, Islamic Rep., Iraq, Israel, Jordan, Kuwait, Lebanon, Oman, Qatar, Saudi Arabia, Syrian Arab Republic, Turkey, West Bank and Gaza, Yemen Rep.
North America	Canada, Mexico, United States
OECD Asia	Australia, Japan, New Caledonia, New Zealand, Korea, Dem. Rep., Pacific island small states, French Polynesia
South America	Argentina, Antigua and Barbuda, Bahamas, The, Belize, Bermuda, Bolivia, Brazil, Barbados, Chile, Colombia, Costa Rica, Caribbean small states, Cuba, Curacao, Cayman Islands, Dominica, Dominican Republic, Ecuador, Guatemala, Guyana, Honduras, Haiti, Jamaica, St. Kitts and Nevis, St. Lucia, St. Martin (French part), Nicaragua, Panama, Peru, Puerto Rico, Paraguay, El Salvador, Sint Maarten (Dutch part), Seychelles, Turks and Caicos Islands, Trinidad and Tobago, Uruguay, St. Vincent and the Grenadines, Venezuela, Virgin Islands (U.S.)

List of Figures

List of Tables

Introduction

1

Between 1950 and 2050, the world population will have increased from 2.5 billion people to more than 9 billion people and global Gross Domestic Product (GDP) will have multiplied by 40 (University of Groningen 2014; The Guardian 2011). Humanity is not in the middle of an evolution. Rather, it is in the center of a complete change of paradigm. In four generations, the world has come to saturation—overpopulated, interconnected, and avidly consuming energy.

Humankind has long lived as a prisoner of its condition. Industrial revolutions freed him from having to depend on nature to provide him with the means to survive. The first industrial revolution was brought about by coal. It created the conditions for the economic rise of Europe and North America. At the beginning of the nineteenth century, China was the world's top-ranked economy in GDP terms (Database BLB 2014). Europe had just started to build machines, roads and railways. Regions began to become accessible, people started to connect on a large scale, trade expanded. Transportation became faster, information exchange sped up, and commercial relationships shaped up throughout the continent. In the second half of the nineteenth century, a long exodus started, with large populations leaving their farms in the countryside for the cities and their factories. Productivity increased considerably. It took one century to complete this evolution. The power of machines created the conditions for incredible achievements and gave these countries a competitive edge. At the beginning of the 1920s, a second revolution started, this time propelled by oil. It started slowly then accelerated during the Second World War, and confirmed the industrial supremacy of the United States. After 1950, the speed of development increased even more. Countries devastated by the war were rebuilt, colonial empires collapsed, new economies emerged. The world then entered a period of relative economic stability. This is the world we live in.

It is the era of easy oil. Despite two oil crises in 1973 and 1979, oil continues to be the primary resource that powers the world economy. It is used to power machines and cars, to produce electricity, and to manufacture plastics and various chemicals which pervade our daily life. A family of four consumes about 800 MJ of

© Springer International Publishing AG 2017
V. Petit, *The Energy Transition*, DOI 10.1007/978-3-319-50292-2_1

energy every day, compared to the 5 MJ strictly necessary to survive (Durand 2007). Every family on earth thus consumes the energy of an army of 160 servants. This is even more striking when one looks at energy consumption in the United States—people there consume an average of 4000 MJ every day, which corresponds to the energy needs of 800 people!

In the last 30 years, new economies have emerged progressively. Countries in Asia—and, soon, Africa—are catching up with the world economy. China and India, the topmost economies at the beginning of the nineteenth century, will reclaim their positions within the next 50 years. The economic world order will then recompose to become similar to the historical balance in place centuries ago. Demographic, territorial and economic powers all reform themselves after a long period of historical apathy. The gaps between mature economies and new economies start to shrink. The technologies which helped some countries to dominate the world's economy are now available to all, allowing weaker economies to catch up. The improvement of sanitary conditions yields an exponential increase in population. New sectors of activity emerge in many countries, each contributing its own multiplier to productivity. The way people live has been transformed, thanks to access to a complete range of products and equipment. With technology, distances start to reduce and old linguistic and cultural barriers collapse. This in turn enables easier deployment of technology, products and equipment to the entire world population, after a transitory era where access to their benefits was strictly limited to those in Europe and North America. The heart of the world finds its way back to Asia.

The current world is also very different from the one of the nineteenth century. Humanity freed itself from enslavement to climatic conditions. The world population now drives the increase of its living standards. Already, massive middle-class populations emerge from populations that were previously conditioned into misery. Many already live by new standards and the cultural differences that once shaped the world and made it diverse and colorful seem to fade away.

Some claim that this evolution is detrimental to humanity. They contend that the improvement of living standards is done using recipes from the past, and that those recipes exhaust the resources of our Earth. With a bigger world population and the homogenization of living standards, the demand for energy has indeed increased in a spectacular manner. For 20 years already, humanity has been consuming more ecological resources than the planet is able to generate (Raisson 2010). Humanity thus needs to somehow define for itself a sustainable future. It needs to transition to a world where it will consume more, but with less.

The energy transition is therefore the real challenge of this century. When future generations look back at our time, they will challenge how we reacted to this transition and radically shaped the world that they live in.

All transitions are complicated by nature. The one we are undergoing is particularly impacted by the number of players and external factors that go into it. Every nation has a responsibility for facilitating it. Every government, too, needs to address it while taking into account its own particularities: the makeup of its economy, its geography and geology, the expectations of its people, and its bilateral relationships. The objective remains always to adapt in order to defend the interests of one's people, the integrity of one's territory, and one's culture and identity. Beyond the borders of each nation, every industry is impacted by this change, whether it concerns the production of primary resources, its transformation or simply the use of energy to build products and services. Every company has to balance sustainable development with competing in a fierce battle to remain attractive to investors. Finally, every action or decision made at one place can create systemic consequences throughout the entire world economy. The energy transition is thus a complex issue; its history remains to be written.

The ambition of this book is to clarify the challenges this transition presents to the world. This is not a book on economics, geopolitics or science. It is more of an overview, because one can only understand the energy transition by combining the different themes into a single perspective. The purpose of this book is indeed to promote and improve understanding of what is at stake and to show what already exists and what can be done to facilitate this transition.

The first part of this book examines the historical continuities that have shaped this transition, the long-term evolutions that developed over several generations. Those long-term evolutions are key. They are the historical determinisms that define the world in which, whatever we do, we will eventually live in. Then, we will look at how the world of energy is responding to the transition now in progress; how oil, coal, natural gas or electricity industries answer the growing needs of the world population. After that, we will explore how these changes could result in various significant geopolitical evolutions. Finally, we will look at how industries and nations adapt to this change.

The world indeed faces a seemingly impossible energy equation. On one hand, it needs more and more energy resources to fuel its development. On the other hand, the way humanity consumes fossil-based energy appears to be leading the planet on the verge of collapse. The energy equation can be solved; that is the object of the last chapter of this book.

References

Database BLB Chine PIB (2014) http://databaseblb.unblog.fr/2007/12/02/evolution-et-chiffres-du-pib-de-1500-a-nos-jours-en-france-royaume-uni-allemagne-etats-uni-japon-chine-et-inde/

Durand B (2007) Energie & Environnement, Les risques et les enjeux d'une crise annoncée. Collection Grenoble Sciences, EDP, Paris

Raisson V (2010) Atlas des Futurs du Monde. Robert Laffont, Paris

The Guardian (2011) http://www.theguardian.com/news/datablog/2011/jan/07/gdp-projections-china-us-uk-brazil

University of Groningen (2014) http://www.ggdc.net/maddison/other_books/new_HS-7.pdf

Historical Determinisms Shaping Tomorrow's World

2

2.1 World Population Explosion

Since the second half of the twentieth century, the world has experienced a radical increase in its population (Planetoscope 2014). This should continue during the first half of the twenty-first century and then stabilize. Within a century, the world population will have thus increased from 2.5 billion people to over nine billion in 2050 (Fig. 2.1).

Within the next 20 years, the world population will increase by more than 1.6 billion people (© OECD/IEA, WEO 2012) to reach over 8.7 billion people in 2035. This disruptive growth comes from the worldwide demographic transition that is still in progress. This transition started with improved access to sanitary conditions by rural populations. As such access became more available worldwide, life expectancy rose and infant mortality dropped, resulting in an exponential increase in world population. Rural families which used to have six to eight children, of which more than half did not reach 20 years of age, saw all their children live to adulthood. Local economies were unable to absorb this massive increase of population. Consequently, youths were forced to seek employment in large cities. Once absorbed into the economy, they started their own families. However, the cost of raising children in cities was (and still is) much higher than that in the countryside. Dual-income families emerged as more women entered the workforce and developed careers. Consequently, the number of children per family progressively dropped as living standards increased. It takes one or two generations to accomplish such a demographic transition. During this time, the population increases significantly. When it is finally over, the population stabilizes again (Larousse 2014), and starts to grow older.

This transition happened in the nineteenth century in Europe and North America during the Industrial Revolution. In the rest of the world, it started during the second half of the twentieth century and will stabilize towards the end of the first half of the twenty-first century. World population should by then reach a peak between 9 and 10 billion people (Geohive 2014) (Fig. 2.2).

© Springer International Publishing AG 2017
V. Petit, *The Energy Transition*, DOI 10.1007/978-3-319-50292-2_2

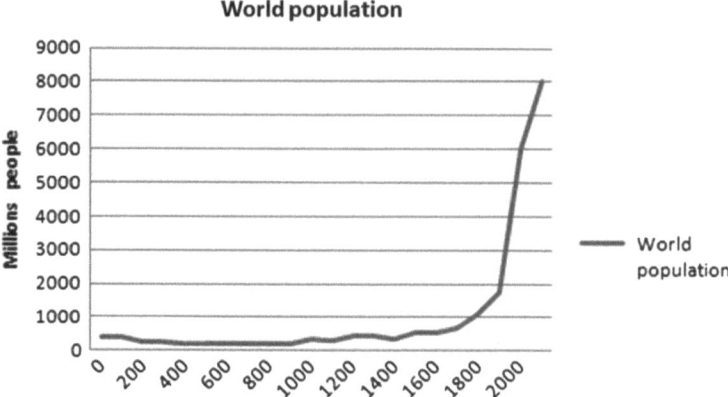

Fig. 2.1 World population (Planetoscope 2014)

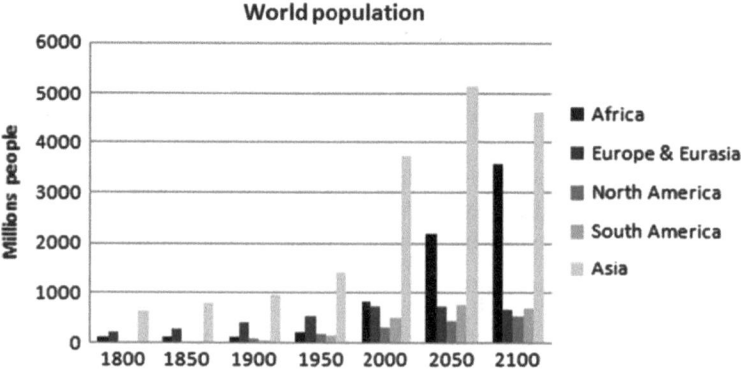

Fig. 2.2 World population (Planetoscope 2014)

In Asia, population growth has started to slow down after 50 years of dynamic expansion. Its population should increase by more than 1.4 billion people by 2050, after an increase of over 2.3 billion people since 1950. It would then reach 5.1 billion people, more than half of the world population. Growth in Africa, however, has just started. Since 1950, the African continent has increased only by 600 million people, but it should increase by more than 1.4 billion people by 2050, and even 2.8 billion by 2100. Africa then follows the same development path, albeit 50 years later than Asia.

As the world population grew, the living conditions of billions of people also improved tremendously. The emergence of a global middle class is actually the turnaround of the twenty-first century. Because of the size of this change, with more people living a new way of life which demands massive quantities of energy, the traditional and precarious energy balance on which the world was built is deeply put in question.

2.2   Rise of New Economies

Asian and African countries will see their population increase by another 2.8 billion people by 2050. This growth is similar to what was experienced in the last 50 years. This demographic transition yields an economic transition, with China in the last 20 years being one of the best examples.

The economic system of new economies is based on massive populations living in rural areas and dependent on primary production (agriculture, fishing, etc.) as their means of subsistence. Such a survival economy is highly dependent upon climatic conditions and external factors. Local industry tends to be of poor quality and essentially serves underdeveloped internal markets. Everything valuable is imported and only accessible to a minority of people. Infrastructure is generally of poor quality, which increases even more the relative isolation in which rural populations survive not just geographically but also culturally, as they remain remote from decision and business centers. The life of these millions of isolated people is highly dependent on external factors that they cannot control. Their world is deterministic.

Access to health services and improved sanitary conditions is the actual change of paradigm. Together with the development of the global economy and increased exchanges between populations, it helps break the isolation of rural populations. Mortality rate decreases, in particular infant mortality. The share of young people in the overall population then increases significantly. However, such a rise is unbearable by the local economy. Consequently, part of the population progressively moves out of these regions towards the cities. This change can be slowed down or accelerated, depending on the particularities of the country's geography or politics. In India, local leaders sometimes forced rural populations out of their village (Das 2000). In Africa, the complete absence of some infrastructure isolates some people for decades and therefore delays this change. Finally, family culture, which encourages (or not) support between generations can accelerate or slow down this movement (Todd and Le Bras 2013). Nevertheless, the change is ineluctable. Large amounts of people press on in the cities to look for jobs. This massive amount of workers is an opportunity for economic growth as their productivity in cities is far higher than in the countryside. The main issue is how to absorb this volume of people in the modern economy at the same speed at which they reach the cities. In China, 8 % of GDP growth is required to allow this integration to happen. The belt of slums that surrounds Indian cities' downtown areas shows that the economic development of the country is not fast enough to absorb displaced populations, despite GDP growth topping 5 % for more than two decades. Then again, the movement cannot be stopped. Growth creates slow absorption so the speed of integration is a major challenge for new economies. It took more than a century for Europe to accomplish it. According to Moshe Lewin (1985), the rise of communism and the terrible events that marked it (like the famines in Ukraine) are simply the consequence of the economic transition that Stalin forced upon the Soviet Union while building a strong industrial economy within a short period of

20 years. Asian and African countries need to realize a similar transition—without the upheaval—within a generation.

Economic development may however take different directions. In China, it led to the creation of gigantic export-oriented industrial powerhouses. In India, it is founded on entrepreneurship and directed towards the domestic market. Cultures vary in essence and the pathways that nations use to fuel their economic development are therefore different. Nevertheless, the result is the same—the economy develops, growth continues, and living standards improve. Increasing state incomes may then be used to improve infrastructure, to break the isolation of remote areas, and to develop the healthcare system as well as education services. Social spending improves the living standards of citizens and the rise of education favors the emergence of the middle class, which drives internal consumption and thus production and economic growth. Growth contributes to pull population even faster out of rural areas, which then needs to be integrated into urban areas, boosting the construction market.

There are of course a number of obstacles to this economic transition. Distribution of revenues can be truncated and accumulated by a minority. Administrative inefficiency can slow down the process, as it did in India in the 1970s and 1980s (Das 2000). Bureaucratic rigidity, endemic corruption and ethnic conflicts can slow down the transition as well, and in some instances give the impression that it has stopped. The complex relationship between rich and poor countries can also hamper the development of new economies because of their dependence on high-value added imports. However, despite these obstacles, be they cultural, economic or political in nature, new economies are bound to succeed simply because it cannot happen differently. A massive middle class will emerge and witness a considerable increase in its living standards. Economic catch-up by new economies is simply a historical determinism. It is inevitable.

Historical trends can only be understood by examining a timeline stretching centuries. Global population movements and exchanges invariably diffuse the benefits from industrial revolutions to the entire world population. Population clusters which remained separated from it will progressively join the world community and eventually benefit from its innovations. Some claim that this transition is harmful to millions of people as they bring about displacements, famines and poverty. This flawed perspective could have come about because such a transition occurs over a long period of time and is difficult to comprehend.

The increase of populations in cities increases urbanization. The urbanization rate worldwide will reach 61 % in 2035 (© OECD/IEA, WEO 2012) and 70 % in 2050 (Le Monde diplomatique 2010). Unsurprisingly, energy consumption in cities is much higher than in the countryside. In addition, the rise of the middle class yields as well a sharp increase in energy consumption. This emerging class can afford to buy personal equipment such as refrigerators, TV sets, cars, etc. which all consume significant amounts of energy. Individual energy consumption has doubled in the last 30 years in Asia, to reach a third of European consumption and a fifth of the American level (EIA 2015) (Figs. 2.3 and 2.4). On the other side of the energy spectrum, almost a fifth of the world population does not have access today

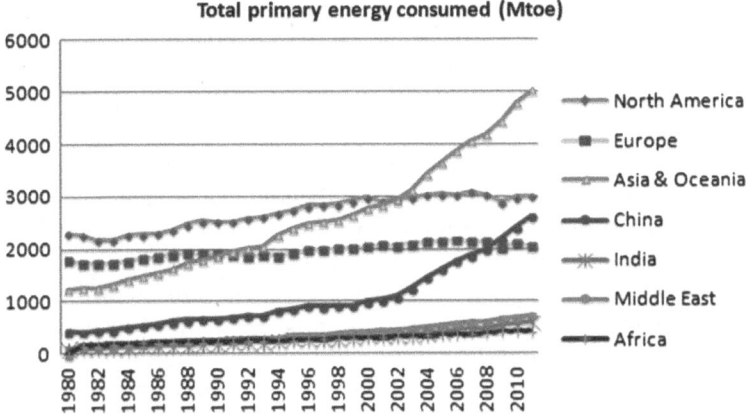

Fig. 2.3 Total primary energy consumed (© OECD/IEA, WEO 2012)

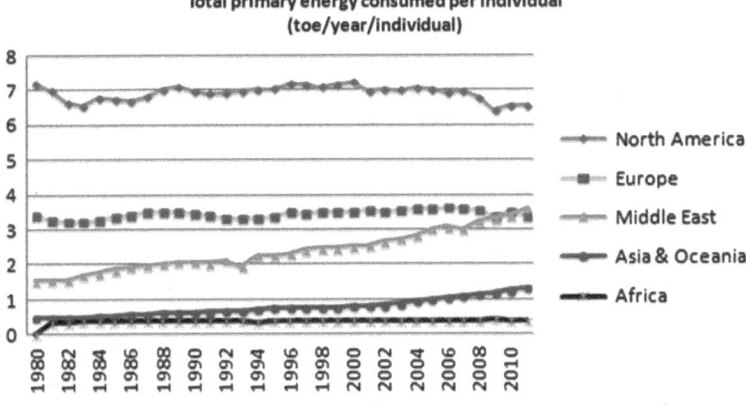

Fig. 2.4 Total primary energy consumed per individual (EIA 2015)

to electricity (© OECD/IEA, WEO 2012), particularly in India, South East Asia and Africa. By 2030, the International Energy Agency plans to reduce this percentage to 12 %, which corresponds, taking into account the concurrent increase in world population, to two additional billion people connected to the electrical network.

If the world population did indeed increase by over 3.5 billion people in the last 50 years [including 800 million for China only (Global Warming 2014)], the individual energy consumption remained stable until the last years of the twentieth century. World population will continue to increase at the same rate in the next 50 years but, on top of that, the energy spent by each individual will rise sharply with rising living standards. Then, energy consumption will skyrocket. This is the main turnaround of the twenty-first century.

Going further, electricity consumption, a subset of overall energy consumption, is the real indicator of economic development. It indeed reflects the consumption of goods and equipment by the middle class. Electricity is the core energy source required to produce equipment and final goods, as well as the energy source for powering these equipment and goods within the new homes of the middle class. According to the International Energy Agency (2012), electricity consumption will grow by more than 70 % in the next 20 years. Electricity consumption today already corresponds to six times the consumption 50 years ago (World Nuclear Association 2014), which means that electricity consumption will have multiplied by more than 10 in a century. This corresponds to an annual average demand growth of 1 % for Europe and 1.3 % for North America, 4.5 % in Asia and 3.7 % in Africa. For Asia and Africa, this growth is considerable.

In a nutshell, economic transition is an ongoing process in most geographies, even though it happens at different speeds and in different conditions. With a massive share of the world population emerging as the middle class, the economic transition yields a sharp increase in energy needs, particularly electricity. Besides being required for producing goods and equipment demanded by the middle class, electricity is also required to power the new houses and fully equipped homes of billions of people. Finally, the greater interlinking of economies, as well as the development of infrastructure, contributes to raise sharply the mobility of people and therefore the consumption that is associated with their transportation needs.

2.3 Energy Use in Industry

As mentioned above, the rise of the middle class in new economies drives industrial production. The interconnectedness of the world economy, also called globalization, contributes to accelerate this phenomenon as it facilitates the access of new economies to equipment and goods manufactured anywhere. The consequence is a sharp rise in industrial production and therefore of energy consumption.

Energy consumption by industry is about 3400 megatons of oil equivalent (Mtoe) per year, about a third of total energy consumption. About 30 % of this amount corresponds to the "energy used by transformation industries and the associated energy losses in converting primary energy into a form that can be used in the final consuming sectors" (© OECD/IEA, WEO 2012). This means that the "final" energy consumption by industry is around 2400 Mtoe (© OECD/IEA Explore 2014), about one third of the total final energy consumption. In this chapter, we shall focus on this "final" energy consumption.

The industry sector in Asia (including China) now represents 50 % of world energy consumption, almost double the 30 % two decades ago. The combined share of Europe and North America, which was 40 % in 1990, has dropped to 28 % today (© OECD/IEA, OECD 2012; © OECD/IEA, Non-OECD 2012; © OECD/IEA, WEO 2012). The significant increase of industrial production in Asia (and its energy consumption) is due to the fast development of Asian economies in recent

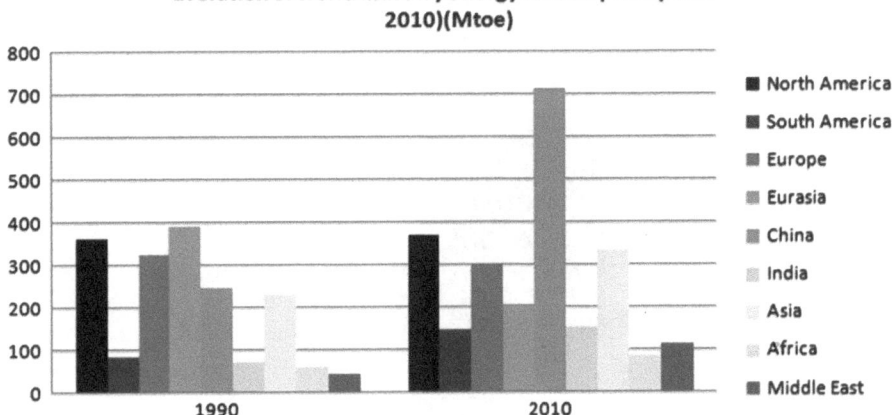

Fig. 2.5 Worldwide industry consumption (© OECD/IEA, OECD 2012; © OECD/IEA, Non-OECD 2012; © OECD/IEA, WEO 2012)

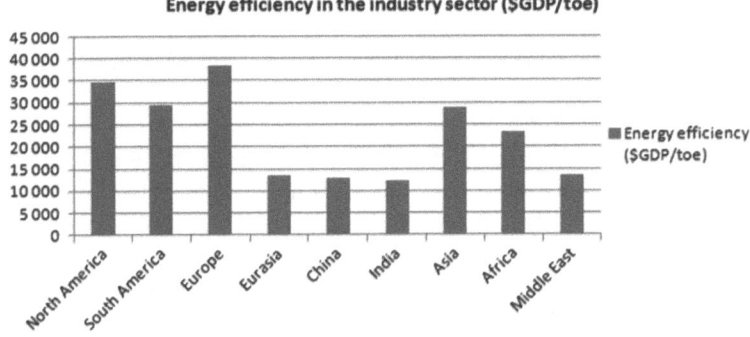

Fig. 2.6 Energy efficiency in industry (© OECD/IEA, OECD 2012; © OECD/IEA, Non-OECD 2012; © OECD/IEA, WEO 2012)

years as well as the globalization of trade. An economic transition is surely in progress (Fig. 2.5).

Energy efficiency can be measured by the volume of GDP generated by a ton of oil equivalent (toe). It varies strongly between geographies. Europe and North America show much higher energy efficiencies than the rest of the world. The Middle East, Eurasia, China and India have the lowest efficiencies. This ratio does not only demonstrate the efficiency of industrial production, but also that energy efficiency varies strongly across industry segments. The energy required to operate a steel plant is of no comparison to that of a machine or a car production plant. As a result, energy efficiency of the different regions is very much linked to the mix of industrial activities of those regions (Fig. 2.6).

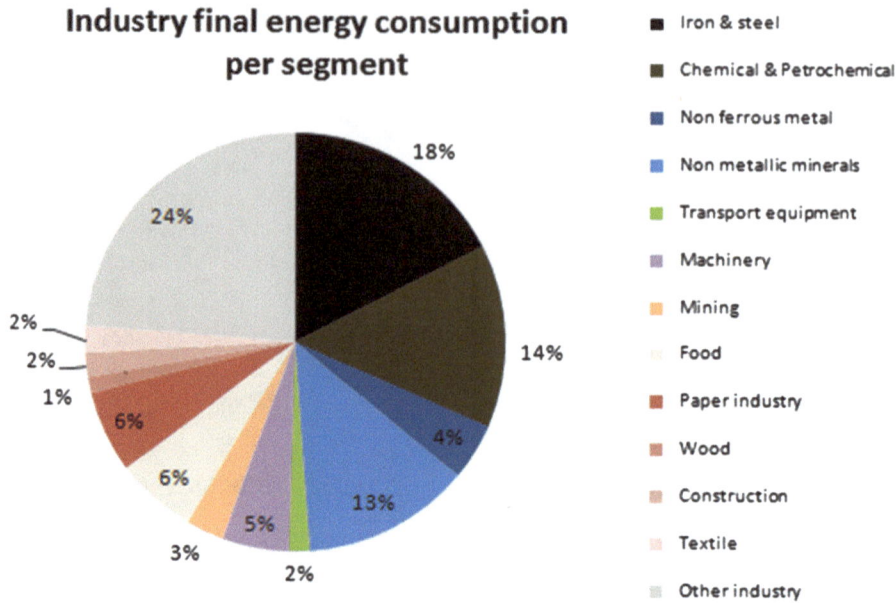

Fig. 2.7 Industry consumption per segment (© OECD/IEA, OECD 2012; © OECD/IEA, Non-OECD 2012)

The main energy-intensive industries are iron & steel, petrochemicals, and mining, metals and minerals (steel, aluminum, various metals, cement, etc.) (© OECD/IEA, OECD 2012; © OECD/IEA, Non-OECD 2012). From an energy standpoint, the mostly energy-intensive operation is the transformation of primary resources to the materials used for the production of goods and equipment (Fig. 2.7).

Almost half of the final energy consumption is used in primary industries such as iron & steel, petrochemicals, and mining, metals and minerals. The steel industry alone represents 18 % of the total final consumption, while petrochemicals consumed 14 % of the total. Other industries are less energy-intensive, but in practice they use the materials and chemicals produced by the transformative industries. Their total energy intensity is therefore dependent on their own consumption of such materials.

The differences in energy efficiency across regions are thus primarily the result of the industrial mix of these geographies. Steel and mineral industries as well as non-metallic minerals industries (cement) dominate in China (© OECD/IEA, OECD 2012; © OECD/IEA, Non-OECD 2012), and account for about half of the associated world energy consumption. The weight of petrochemicals-related energy consumption is more balanced across regions. Regions with high consumption of oil-related products such as China, North America and Europe have a higher number of facilities and therefore sizeable shares of world energy consumption (Fig. 2.8).

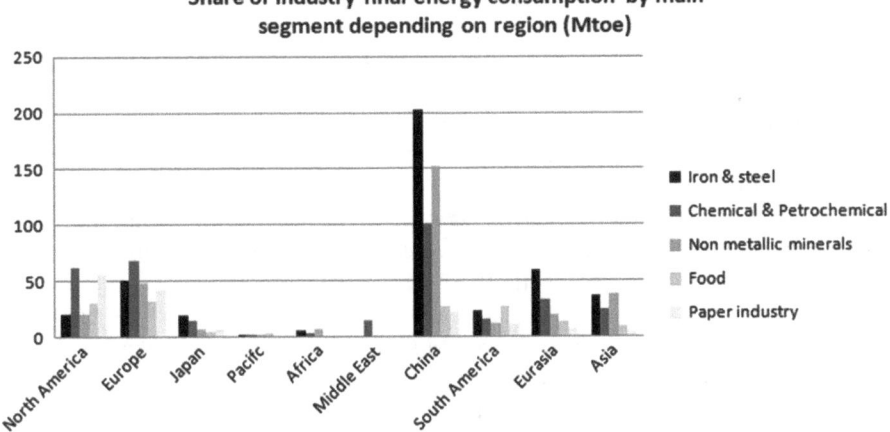

Fig. 2.8 Industry consumption per segment & region (© OECD/IEA, OECD 2012; © OECD/IEA, Non-OECD 2012)

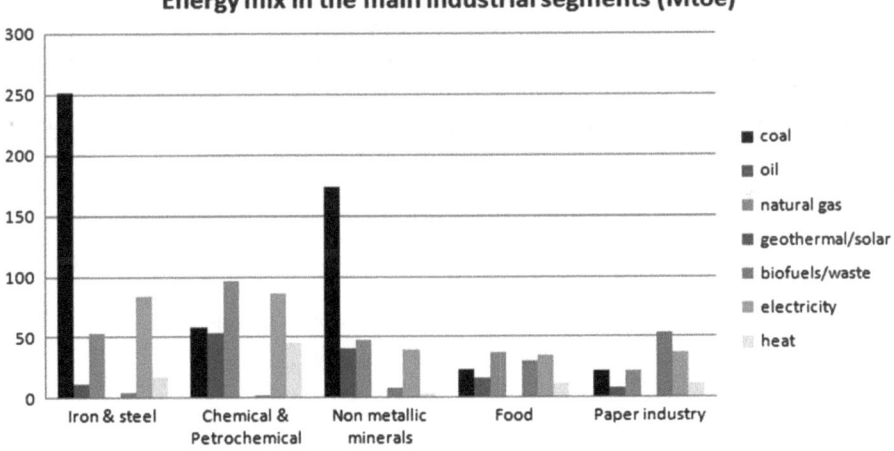

Fig. 2.9 Energy mix per segment (© OECD/IEA, OECD 2012; © OECD/IEA, Non-OECD 2012)

Energy consumption by industry comes in main part from fossil resources, at about 63 % of the total. This share is even stronger when one considers that the electricity used in the various industrial processes is produced essentially from fossil fuels (coal and gas).

The steel segment is highly dependent upon coal (60 % of total primary resources' consumption); the chemicals and petrochemicals industry uses 30 % of natural gas, and 15 % each of coal and oil (Fig. 2.9).

The development of new economies will bring about a significant increase in industrial activity and therefore of energy consumption (© OECD/IEA, OECD 2012; © OECD/IEA, Non-OECD 2012). The International Energy Agency estimates that it could increase by as much as 45 % in the next 25 years, which corresponds to an annualized growth of around 1.5 %. The transformative industries' energy consumption is expected to increase at the same time by 24 % to reach 1700 Mtoe, or 33 % of the total industry sector's energy consumption. All in, the total energy consumption of industry as a whole is set to grow by around 35 %. Other sources present similar forecasts around 32–35 % of growth in the coming decades (Exxon Mobil 2016; Shell 2016) (Fig. 2.10).

As already explained, Asia's share of total industrial production has risen sharply in the last 20 years while the combined share of Europe and North America has dropped over the same period. This trend will continue during the next two decades, as Asia becomes the "workshop of the world". Consequently, 73 % of the energy demand in industry shall come from Asia in the next 20 years, with the rest of the demand coming from Latin America, Middle East and Africa. These three regions have rising energy needs, and their competitive positioning will improve while living standards and therefore the cost of living increase in Asia, in particular China.

Primary resources consumption is thus set to increase, even though the energy mix will change. The share of coal will slowly decline (although the demand will increase in absolute value) as it is partially replaced by electricity and gas in heat production processes. The demand for electricity and natural gas will grow significantly in the next 20 years. Also, natural gas will grow its share in overall electricity production vis-à-vis coal (Fig. 2.11).

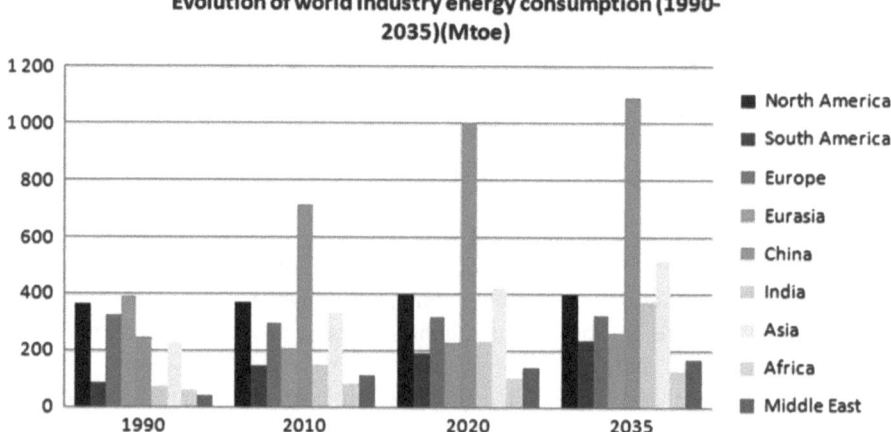

Fig. 2.10 Evolution of industry energy consumption (© OECD/IEA, OECD 2012; © OECD/IEA, Non-OECD 2012; © OECD/IEA, WEO 2012)

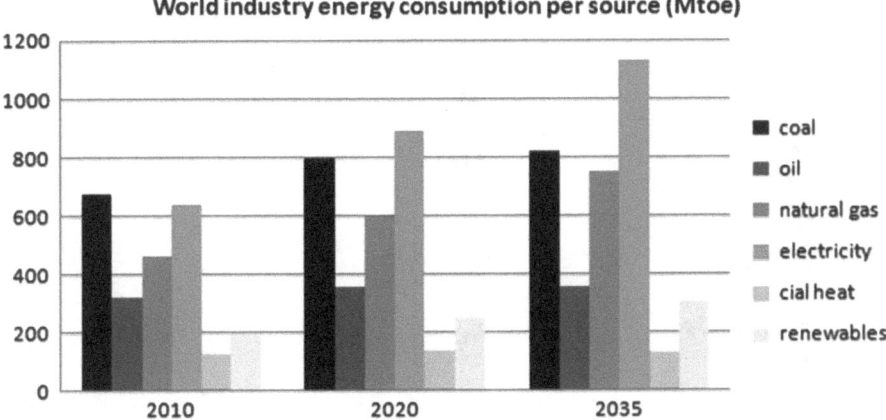

Fig. 2.11 Evolution of industry consumption per source (© OECD/IEA, OECD 2012; © OECD/IEA, Non-OECD 2012; © OECD/IEA, WEO 2012)

Rising living standards lead to sharply increasing industrial activity worldwide. Trade globalization engendered the displacement of industrial production to Asian countries (particularly China), which consequently developed a very large base for industrial production and know-how, and which also allowed the absorption of rural populations by cities and therefore the progressive emergence of a middle class. Consequently, industrial production increased considerably. The associated energy consumption is expected to rise by another 45 % within the next 20 years (or 35 % if we include transformative industries). This sharp increase will likely occur mainly in Asia, which will represent 73 % of the total increase. The energy mix of coal, gas, oil, and electricity will transform progressively towards a higher share of electricity and natural gas versus a declining share of coal and oil.

2.4 Energy Use in Buildings

Everywhere on the planet, television and cinema broadcast the appealing image of a modern way of life to even the most remote areas. When the middle class achieves wealth and comfort, it naturally looks to the services and products now available to them: refrigerators, washing machines, TV sets, the Internet, cars, air conditioning, etc. This naturally brings about a considerable increase in energy demand.

In 2010 the energy consumption in buildings was 2800 Mtoe (© OECD/IEA, Buildings 2013), or about 40 % of total final energy consumption. Residential buildings accounted for 75 % of this amount, and the tertiary segment the rest. The tertiary segment represents here all commercial and industrial buildings which host the various service businesses. Countries from the OECD, while having only 18 % of the world population, accounted for 44 % of the total consumption. World energy consumption in buildings grew by more than 30 % from 1990 to 2010 (Fig. 2.12).

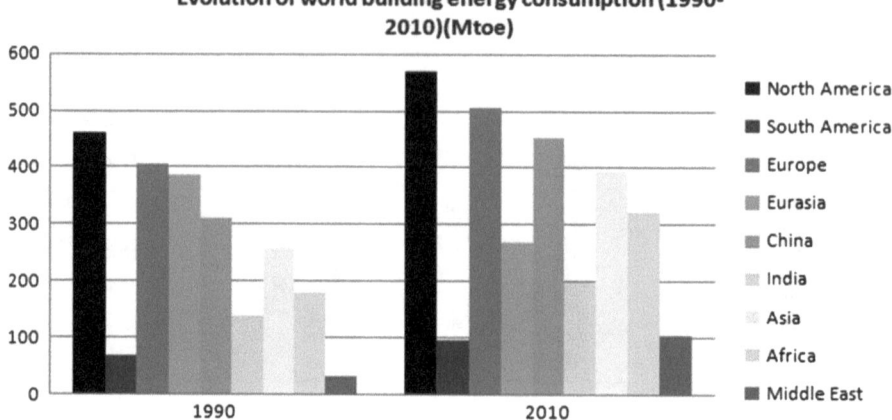

Fig. 2.12 Worldwide buildings consumption (© OECD/IEA, OECD 2012; © OECD/IEA, Non-OECD 2012; © OECD/IEA, WEO 2012)

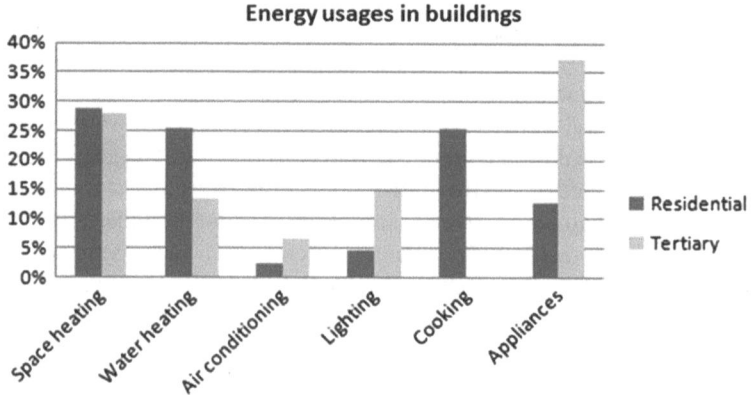

Fig. 2.13 Energy usages in buildings (© OECD/IEA, Buildings 2013; © OECD/IEA, WEO 2012)

Energy consumption in buildings is made up of a number of elements which, depending on the usage and occupancy conditions, may vary (© OECD/IEA, Buildings 2013). The primary use of energy in buildings is for heating space and water. This is followed by lighting and cooking, and then appliances such as TV sets, refrigerators, washing machines, the Internet, etc.

The share of appliances and lighting is much higher in the tertiary segment than in the residential segment; water heating and cooking consume more energy in residential buildings than in tertiary ones (Fig. 2.13).

Energy consumption profiles vary across regions. In the residential segment, individual consumption is much higher in OECD countries, with an average of

0.6 toe/year/individual. This is two times more than the world average of about 0.3 toe/year/individual. Generally, much more energy is used for heating in OECD countries, which can be explained by the fact that most of these countries are located in geographies where climatic conditions are cold. Energy consumption by appliances in OECD countries is also much higher than in other countries (Fig. 2.14).

Urbanization is a key factor in energy consumption in buildings. There is a direct relationship between urbanization rate and energy use. Some exceptions do exist, like in South Africa or Mexico. This is related to the fact that those are still economies in transition. The urbanization rate in these economies has evolved faster than the emergence of the middle class and therefore the consumption of energy (Fig. 2.15).

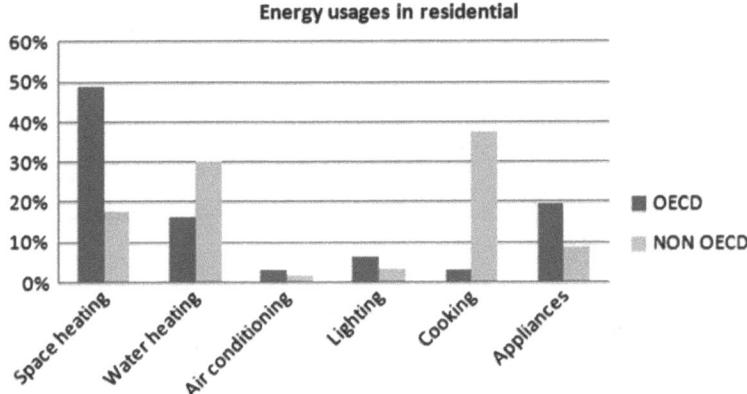

Fig. 2.14 Energy usages in residential (© OECD/IEA, Buildings 2013; © OECD/IEA, WEO 2012)

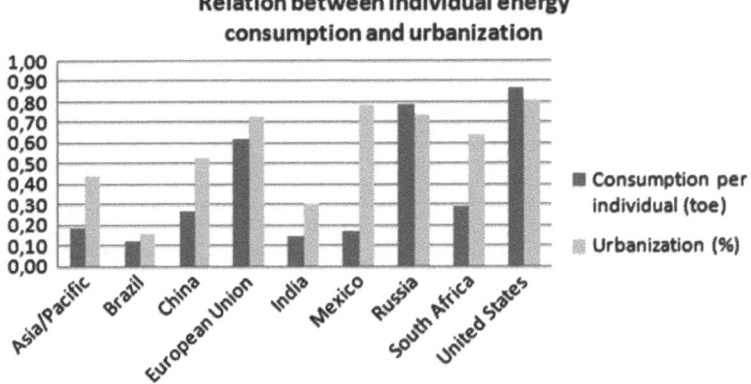

Fig. 2.15 Energy vs Urbanization in buildings (© OECD/IEA, Buildings 2013; © OECD/IEA, WEO 2012)

Energy consumption is not only driven by climatic conditions or by the size of the middle class. It is also related to living habits. For instance, the residential energy intensity in the United States (0.86 toe/year/individual) is much higher than in Europe (0.62 toe/year/individual). The main difference is related to the energy consumption of appliances inside homes. Also, heating consumes less energy in the United States than in Europe (Fig. 2.16).

In the rest of the world, China has a specific situation as winters in many parts of its geography are harsh, while India has one of the lowest urbanization rates in the world (31 %) and therefore limited access to energy (Fig. 2.17).

In summary, energy consumption by individuals in the residential segment is much lower in non-OECD countries. The consumption profiles are also different as

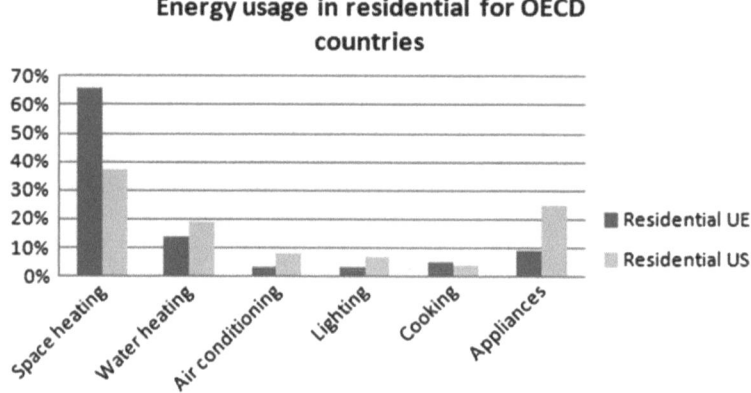

Fig. 2.16 Energy usage in residential in OECD (© OECD/IEA, Buildings 2013; © OECD/IEA, WEO 2012)

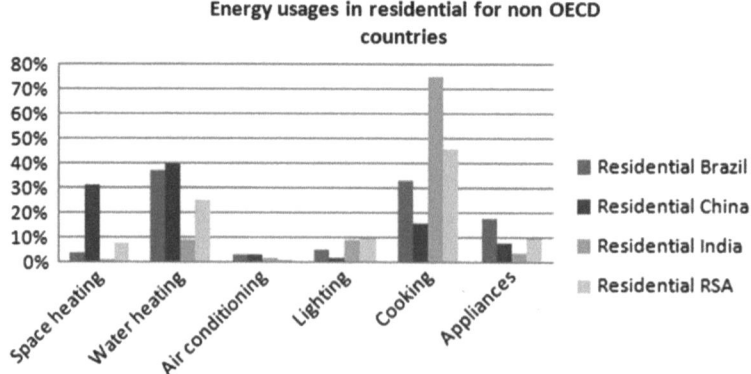

Fig. 2.17 Energy usage in residential in non OECD countries (© OECD/IEA, Buildings 2013; © OECD/IEA, WEO 2012)

the share of appliances (living standards) and the share of heating (climatic conditions) impact strongly the overall consumption mix.

The tertiary segment is much more homogeneous. This indicates some standardization across the globe with regards the use of energy within tertiary buildings (Fig. 2.18).

Increasing urbanization as well as improvement in living standards will mechanically lead to higher energy consumption in buildings. According to the International Energy Agency (2012), energy consumption in buildings worldwide is set to increase by another 30 % by 2035 and by up to 50 % by 2050 (© OECD/IEA, Buildings 2013) (Fig. 2.19).

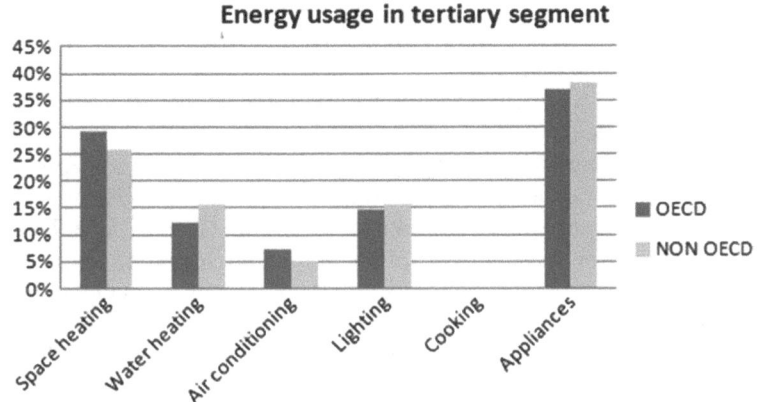

Fig. 2.18 Energy usage in residential in tertiary buildings (© OECD/IEA, Buildings 2013; © OECD/IEA, WEO 2012)

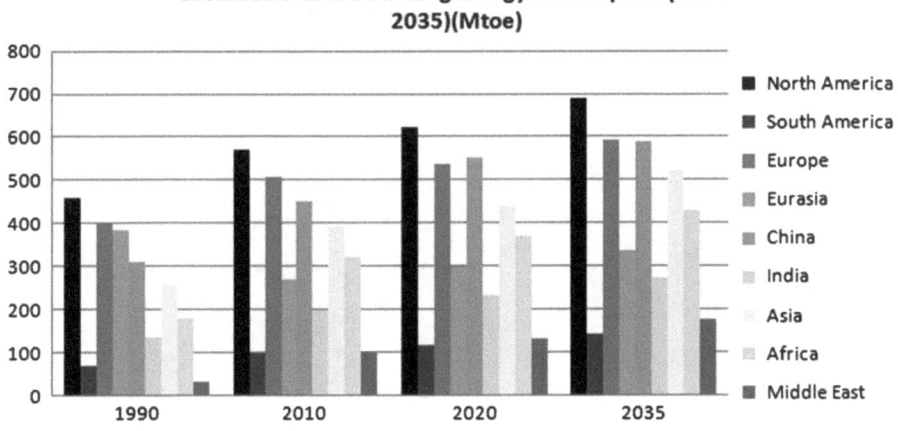

Fig. 2.19 Evolution of buildings' energy consumption (© OECD/IEA, WEO 2012)

Different scenarios exist, however, depending on the energy efficiency measures that could be applied (© OECD/IEA, Buildings 2013). The "6-degrees scenario" (6DS) assumes no specific action to tame energy consumption will be taken. The International Energy Agency estimates that this scenario would lead to 6 degrees temperature increase at the surface of the planet by the end of the century, hence its name. The "2-degrees scenario" (2DS) considers measures to limit the energy footprint of buildings will be undertaken, reducing the impact on planet's temperature to two degrees. The 6DS scenario considers an annualized growth of consumption of 0.7 % in the residential segment and 1.5 % in the tertiary segment; the 2DS scenario assumes zero growth in the residential segment and around 0.7 % growth in the tertiary segment. Other sources such as Exxon Mobil (2016) and Shell (2016) present similar forecasts as the 6DS scenario for the coming years. Exxon Mobil presents an overall growth of the sector of 0.93 % per year. Shell presents a growth of 0.7 % per year in the residential segment, and up to 1.9 % per year in the tertiary segment (Figs. 2.20 and 2.21).

In the 6DS scenario, consumption per individual will rise up slightly to 0.43 toe/year/individual (vs 0.4 toe/year/individual today). This rise is driven by economic development and is largely compensated by the modernization of buildings in a number of areas, which brings about improved energy efficiency. The 2DS scenario assumes that important measures are taken to tame the increase in energy consumption, particularly in the residential segment. The world consumption would thereof reach an average of 0.33 toe/year/individual by 2050, a 20 % drop from today.

The mix of energies used in buildings will also vary strongly in the coming years, towards a greater mix of natural gas and electricity (Fig. 2.22).

Whatever the scenario, the usages of energy and consumption profiles will evolve strongly in the years to come. Energy efficiency and insulation of buildings will lead to decreased energy waste in space heating, but this would be largely

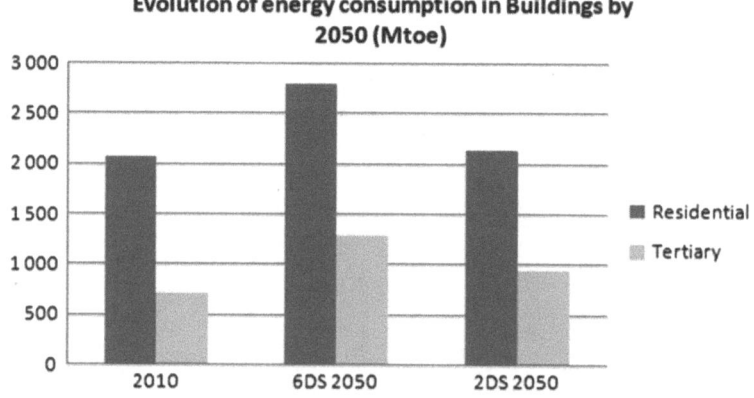

Fig. 2.20 Evolution of buildings' energy consumption per segment (© OECD/IEA, Buildings 2013)

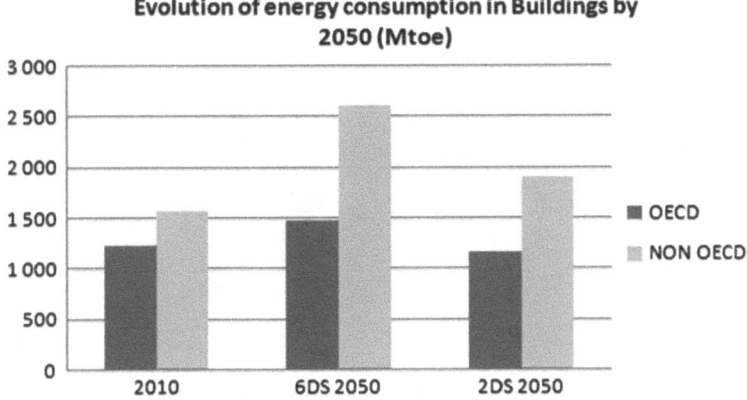

Fig. 2.21 Evolution of buildings' energy consumption per region (© OECD/IEA, Buildings 2013)

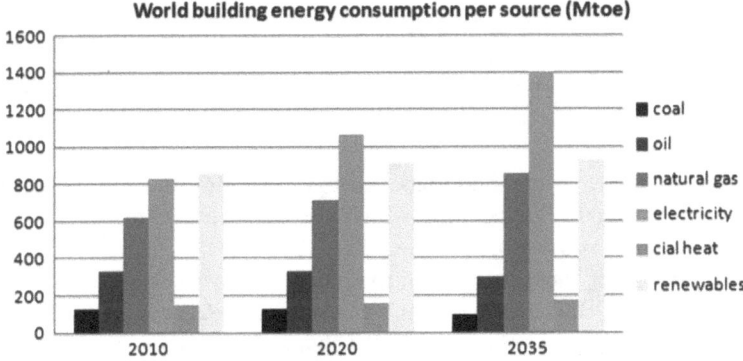

Fig. 2.22 Evolution of buildings' energy consumption per source (© OECD/IEA, Buildings 2013; © OECD/IEA, WEO 2012)

compensated by the rise in energy consumption by electrical appliances. The development of information technology, notably in OECD countries, would lead to more energy consumption by appliances, especially as more get connected to the Internet. In other regions of the world, usages are expected to evolve more slowly (Fig. 2.23).

In the tertiary segment, the consumption profiles should not evolve significantly.

In summary, world energy consumption is expected to increase about 1.5 % on average in the tertiary segment and around 0.7 % per year in the residential segment. This growth is linked to the increase in the world population but, more importantly, to improvement in living standards. Urbanization rate will increase by ten points in the next 20 years, driving higher energy consumption.

This growth will be partially offset by the progressive renovation of old energy-inefficient buildings, particularly in OECD countries. These renovations present an

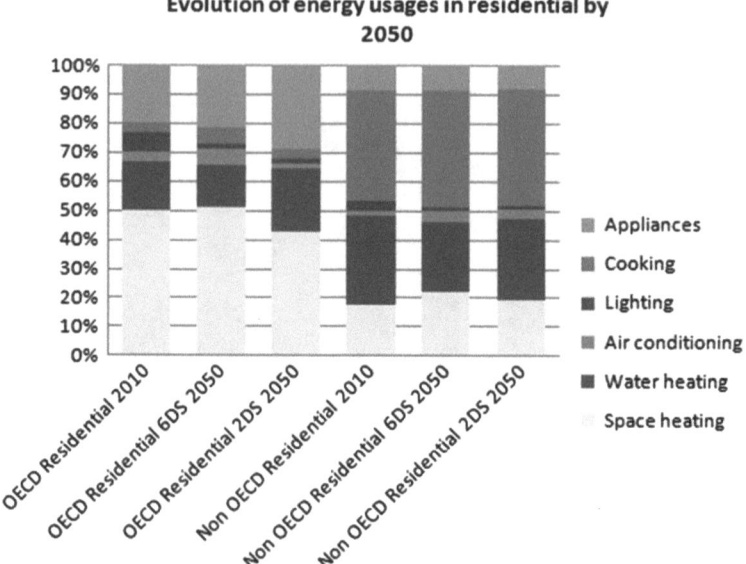

Fig. 2.23 Evolution of buildings' energy usages (© OECD/IEA, Buildings 2013; © OECD/IEA, WEO 2012)

opportunity to improve the buildings' energy footprint and therefore to lower consumption, especially for heating, which share in OECD countries is expected to drop by as much as five points in the next 20 years.

At the same time, the pervasive introduction of new technologies in the residential segment will lead to increased energy usage in OECD countries. According to the International Energy Agency (2012), the share of energy consumed by electrical powered appliances in homes will increase by up to nine points in OECD countries.

2.4.1 The Boom of Information and Communication Technologies

Nowadays more electrical appliances and devices can connect with one another and to the Internet. This trend will continue and lead to a rise in the share of energy consumed by such machines. According to the Bank of America (Financial Times 2014), the information and communication technologies (ICTs) segment accounted for 10 % of worldwide electricity consumption. The segment's consumption has increased considerably over the last 10 years. Data centers alone represent between 1.1 and 1.5 % of worldwide electricity production (Koomey 2014). Consumption doubled between 2000 and 2005 and grew 56 % between 2005 and 2010, and should continue to grow rapidly! The number of connected devices reached 12 billion in 2010 and should increase 25–50 billion by 2020 (Cisco 2011; FierceWireless 2015). The deployment of ICTs throughout the world will therefore bring about a considerable increase in the electricity consumption by the segment. The boom of new technologies is therefore also a historical determinism which will significantly contribute to the energy transition.

2.5 Energy Use in Transportation

Mobility is probably one of the most important boosters of economic development and of the progress of humanity in general. The interconnection of cultures, languages and economies allows the sharing of ideas, know-how, products, and financing. It is this connectivity which led to the rise of economies in North America and Europe in the nineteenth century. Railroad development put an end to the historical isolation of many populations and helped value differently large portions of underdeveloped territory.

With mobility comes wealth. The development of infrastructures is therefore essential to allow mobility to progress. One of the primary roles of governments across the world is to provide people with the means to better value their territory. Beyond infrastructure, cars, buses, trains, airplanes transport people from one place to another. Transportation usage is directly linked to the economic development, and transportation is an important energy consumer in the overall energy mix.

Transportation today corresponds to 28 % of total energy consumption, with more than 2000 Mtoe (© OECD/IEA, Explore 2014). The figure excludes air and marine bunkers (around 350 Mtoe in 2010; © OECD/IEA, Statistics 2015). Energy consumption in the transportation sector has grown by almost 50 % in 20 years; at the same time, the efficiency of engines and other devices has considerably increased (Fig. 2.24).

There are a multitude of usages in transportation. Road transportation is by far the most popular, accounting for 64 % of total transport in kilometer terms (© OECD/IEA, Transport 2009), and over 75 % of the associated energy consumption. Long-distance transportation represents between 10 and 20 % of the total

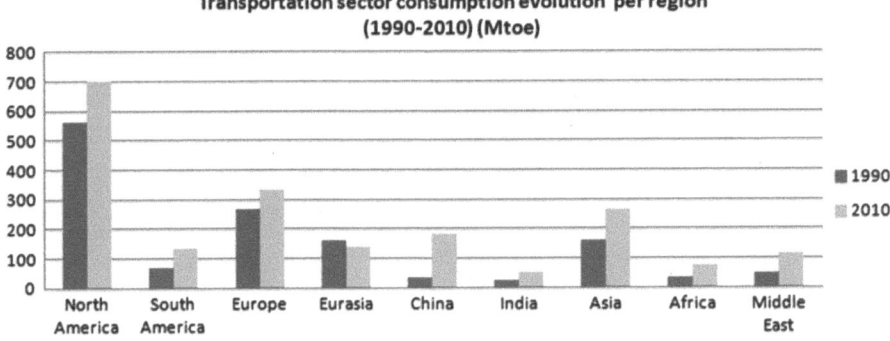

Fig. 2.24 Worldwide transportation consumption (© OECD/IEA, Explore 2014; © OECD/IEA, Statistics 2015; © OECD/IEA, WEO 2012)

transportation in most countries, except in Asia (excluding China & India) and in Africa. The small share of long-distance transportation there, around 6 %, limits economic development. Not surprisingly, the type of transport used for long-distance travels also varies across regions. Air transportation is favored in OECD countries, while rail transport remains the primary mode in China and India.

Short and medium distances remain the primary type of travel, making up around 80 % of the total, and up to 94 % in Asia. In OECD countries, the use of private cars dominates. In the rest of the world, collective transportation is more developed, and in Asia two-wheeled transportation (bicycles, motorcycles) form an important share of the market (Fig. 2.25).

Total mobility is measured in kilometers. That in OECD countries (25 trillion kilometers) equaled the rest of the world in 2005. However, by 2050, mobility in OECD countries should rise up to 30 and 80 trillion kilometers elsewhere (© OECD/IEA, Transport 2009). This sharp increase in new economies is due to both the increase in the world population as well as the increase in mobility per individual.

Most of the world population increase comes from non-OECD countries (Geohive 2014) and therefore the increase of mobility in absolute value is first linked to this rise. The mobility by individual in OECD countries should slowly increase to 20,000 km/year/individual (against 17,000 km/year/individual today), while the mobility in the rest of the world should double from 5000 km/year/individual to 10,000 km/year/individual (Fig. 2.26).

Consequently, there should be a significant increase in energy consumption in the sector. This consumption is set to increase by another 40 % within the next 20 years. Different sources converge towards this level of growth (© OECD/IEA 2012; Exxon Mobil 2016; Shell 2016). The growth will mainly be in Asia (including China and India), which should represent 75 % of the total growth of the sector (Fig. 2.27).

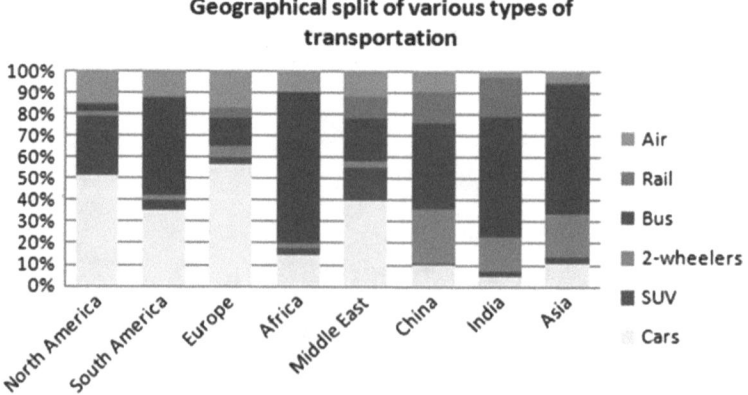

Fig. 2.25 Transportation usages (© OECD/IEA, Explore 2014; © OECD/IEA, Transport 2009)

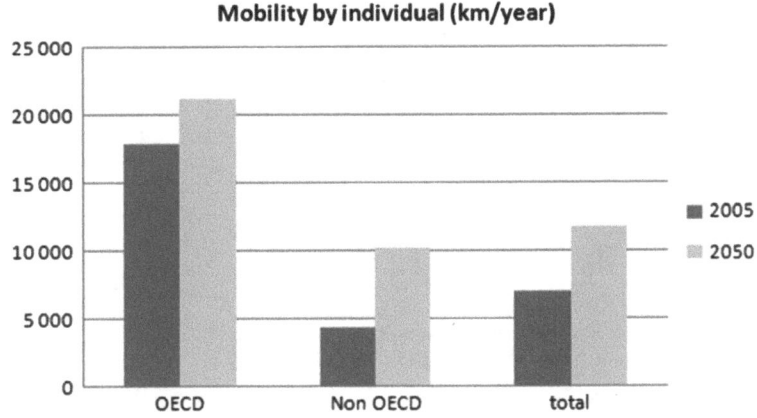

Fig. 2.26 Mobility by individual (© OECD/IEA, Transport 2009)

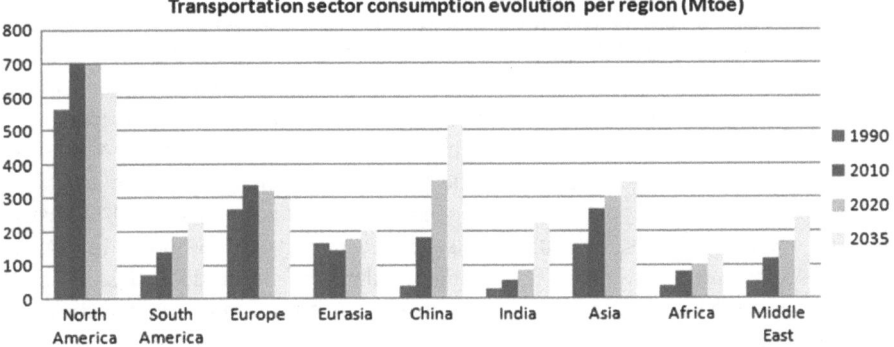

Fig. 2.27 Evolution of transportation consumption (© OECD/IEA, Explore 2014; © OECD/IEA, Transport 2009; © OECD/IEA, WEO 2012)

This sharp increase of mobility in non-OECD countries will be driven primarily by light road (cars, two-wheelers, etc.) and long-distance transportation (notably air transportation); they should each represent 40 % of the total growth. Asia is already the primary growth market for air transportation, and Africa should develop quickly in the years to come (Fig. 2.28).

In summary, the economic development associated with the interconnection between different regions of the world creates the conditions for a strong rise of mobility in new economies. Mobility will increase as people get more interconnected in the global economy. It is consequently expected to double in the coming decades, with the associated energy consumption to increase by 40 % by 2035.

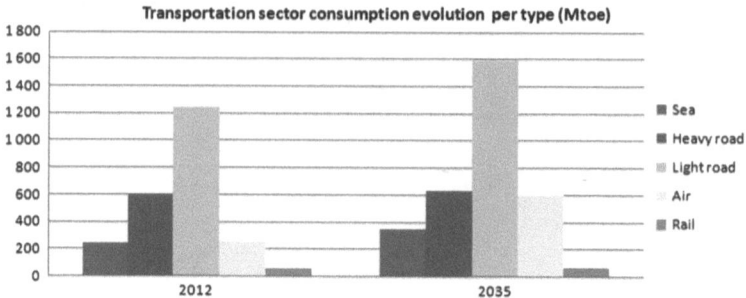

Fig. 2.28 Evolution of transportation consumption per segment (© OECD/IEA, Explore 2014; © OECD/IEA, Transport 2009; © OECD/IEA, WEO 2012)

2.6   Impact on Climate Change

Within the next 50 years, the world population will have completed its demographic transition, moving from 2.5 billion people (in 1950) to more than 9 billion in 2050. Then, it will reach a plateau, after which it will slightly decrease. Meanwhile, globalization will have spread its benefits across new economies. New economies will have emerged and will have caught up with mature economies. A massive middle class will emerge, with needs and wishes that correspond to those of their counterpart in OECD countries. Consequently, the industry sector will adjust to provide for these needs, causing the associated energy consumption to rise by around 45 %, mostly in Asia. The living standards of the middle class will see much improvement, leading to a surge in energy consumption in buildings. If nothing is done, the increase of energy consumption in buildings in new economies should be three times faster than in mature economies, and increase overall by around 30 %. Finally, globalization will connect billions of people and give them the ability to integrate into the global economy. Consequently, the mobility of this emerging middle class will increase considerably. The total mobility in the world shall more than double in the next four decades. As a consequence the energy consumption associated with mobility is expected to grow by 40 %. In the end, if nothing is done, the world will consume 35 % more primary energy in 2035 than today, and up to 50 % more by 2050.

Scientists from the Intergovernmental Panel for Climate Change (IPCC) (2007) have measured the increase of temperature on Earth. They demonstrated that the temperature rose by 0.74 degrees on average between 1906 and 2005; the rise was only 0.6 degrees between 1901 and 2000. As well, sea levels have risen on average by 1.8 mm per year since 1960, and 3.1 mm per year since 1993. Their report confirms both the increase and the acceleration.

Many different factors can influence the increase of the temperature at the surface of the planet. First, astronomic evolutions have an important impact on the climate on Earth as they lead to evolutions in radiative emissions from the sun.

The Serbian climatologist Milutin Milankovitch (1920) identified three main astronomic origins of Earth's climate evolution: orbit eccentricity, the regular evolution of the elliptic shape of Earth's orbit around the sun; obliquity, the variation of the angle between the rotation axis of the planet and an axis perpendicular to its orbit; and precession, the evolution of the planet's rotation itself. These astronomic variations are mainly linked to gravitational attraction evolutions between Earth and other planets. Beyond these evolutions, changes in solar activity and sun spots also have an important impact. On the planet, the movements of tectonic plates our continents sit on also contribute to climate change as they modify ocean currents. Volcanic activity also has a non-negligible impact.

Another factor in climate change is human activity and its impact on greenhouse gas emissions, also called "anthropogenic forcing". Earth receives every second around 340 W/m^2 of radiation from the sun. A part of this radiation remains trapped inside the atmosphere (around 100 W/m^2). A higher greenhouse gas concentration in the atmosphere makes it more difficult for this trapped radiation to leave the atmosphere. A larger share remains trapped and contributes to an increase in Earth's surface temperature. "Anthropogenic forcing" today stands at 2 W/m^2, which corresponds to around 400 ppm of CO_2 concentration (largest contributor before methane, around 75 % of total emissions), a level which never exceeded 280 ppm in the last 650,000 years (Durand 2007).

All these factors need to be taken into account when building a climate model. The vast majority of scientists today recognize that the recent acceleration of climate warming can only be explained by "anthropogenic forcing". Many national science academies across the world have publicly acknowledged this (National Academies 2009). Now, a small group of experts continue to question this as they challenge the accuracy of current climate models. They however recognize unanimously that the increase of greenhouse gas concentration in the atmosphere will lead eventually to climate evolutions.

Greenhouse gases have different atmospheric lifetimes. Methane stays on average 10 years in the atmosphere; it takes 50–200 years for CO2 to disappear. Daily emissions of methane and CO2 thus accumulate in the atmosphere, accelerating the phenomenon. Greenhouse gas concentration has increased by more than 40 % in a century (OMM 2014) and the daily emissions have increased by more than 70 % between 1970 and 2004 (GIEC/IPCC 2007). The concentration phenomenon thus accelerates; it is a historical determinism.

If we focus on CO_2 emissions, six regions represent more than 68 % of the world emissions, and three of them (North America, Europe and China) represent half of the total. The United States represents 25 % of the world's CO_2 emissions with only 4 % of the world population. Europe represents 14 % of total emissions (© OECD/ IEA CO2 2013). China and India still have lower emissions, but they are far from having realized their economic transition, and the volume of their emissions grows quickly. Their share shall increase significantly in the coming decades (Fig. 2.29).

A full 40 % of CO_2 emissions come from the production of electricity, and a bit less than a quarter from transportation. Any ambition to stabilize the volume of emissions will then be related to efforts in those two domains (Fig. 2.30).

Geographical perspective of CO2 emissions

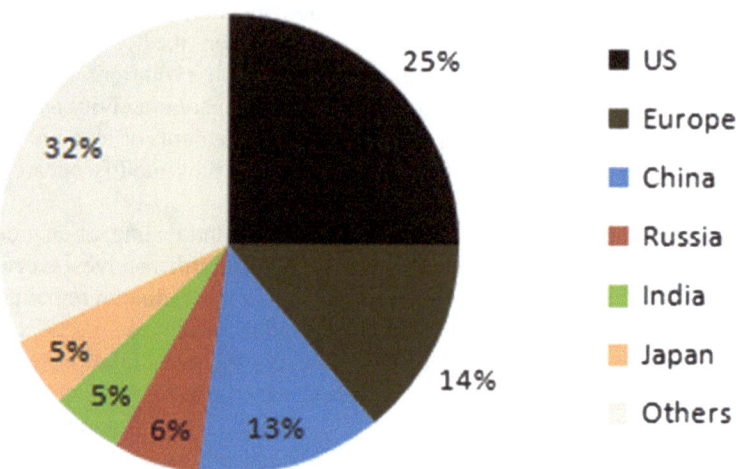

Fig. 2.29 CO_2 emissions per region (GIEC/IPCC 2007; © OECD/IEA, CO2 2013)

CO2 emissions by origin

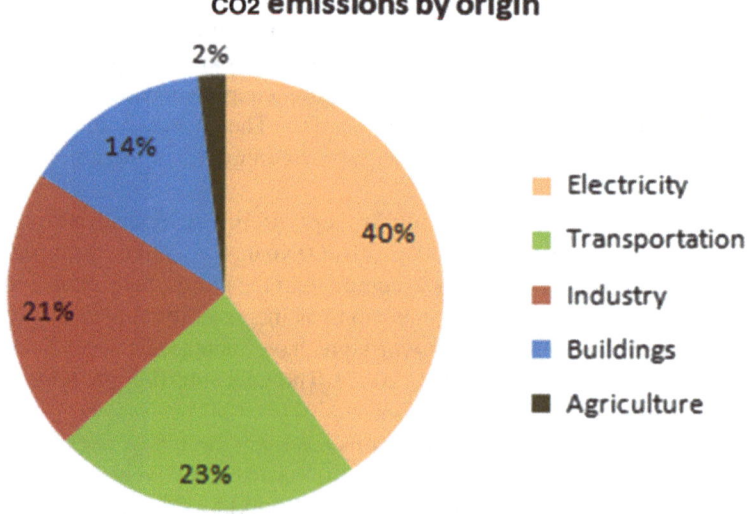

Fig. 2.30 CO_2 emissions by origin (GIEC/IPCC 2007; © OECD/IEA, CO2 2013)

The IPCC has come up with a number of scenarios to predict the evolution of greenhouse gas concentration and its impact on the planet's climate. They do not measure any probability of occurrence but consolidate a vast array of indicators, such as world population growth, economic growth, evolution of multilateral government relationships, political stability, and possible energy supply choices. These different scenarios foresee an increase in Earth's temperature of between two and four degrees by the end of the century. Even if greenhouse gas emissions were limited to their pre-industrial level (obviously impossible), the current concentration would lead to a temperature increase of at least 0.6 degrees by the end of the century.

A number of consequences are thus expected. The IPCC (2007) ranks them between "probable" and "highly probable". The modifications of the structure of snowpack, the increase of the size of mountain lakes, and the intensification of runoffs would be among the primary observable consequences of temperature increase. Cyclonic activity, particularly in the Atlantic Ocean, should increase as well. Sea levels would rise and the oceans would become more acidic, modifying the cycles of rain. Wind and marine streams, linked to temperature gradients across regions, would also be modified. This would have an impact on the world's climate as we know it. Rainfall should be more frequent at higher latitudes and rarer at lower latitudes. This would in turn bring about consequences on ecological balance in those regions as well as on the way people live. Seasons should be modified, as well as yearly migrations of a number of animal species.

Fragile ecological systems would first be threatened. The acceleration of climate evolution is indeed the major threat to these ecosystems because of the lack of time to adapt to these changes. Mountains, Mediterranean regions, tropical forests and marine systems such as coral reefs would be among the most impacted. Poor populations would also be impacted first, notably in central Africa, with the rise of water stress and the transformation of forests into savanna, or with the rise of sea level in some parts of Asia.

Finally, consequences could be more serious. Beyond a temperature rise of two degrees, 30 % of the living animal species could disappear, considerably modifying the food chain on Earth. Beyond four degrees, these extinctions could be massive. With a three-degree rise, 30 % of coastal areas could be submerged by oceans, and a number of pathogens could migrate from their original region to others.

The world seems to have acknowledged for a few years the terrible consequences on the climate of our way of living. This is partly thanks to the work and efforts of the IPCC. Still, despite numerous international climate summits, a global response has yet to be defined. The Kyoto protocol (not ratified by the United States) led to a number of improvements as many countries adopted specific policies and targets on their emissions, but the execution of those plans proved to be complex as it can impact the competitiveness of nations. More recently, the COP21 conference on climate change (December 2015), held in Paris, was an important step forward in the right direction, although the agreement remains to be ratified by all countries. As the world moves on, the historical catch-up of new economies remains anyhow the primary objective, and international competitiveness continues to be the primary indicator. While this happens, energy consumption continues to rise and greenhouse gas emissions accumulate in the atmosphere. The impact of human activities thus continues to worsen.

2.7   Summary

By 2050 the world population will have reached its peak of around 9–10 billion people, 2.5 billion more than today and almost four times the population a century ago. Add to this peak population new economies and increasing urbanization, and we end up with a sharp increase in energy consumption in general and electricity consumption in particular. According to the International Energy Agency, primary energy consumption could increase by 35 % in the coming 20 years and by up to 50 % by 2050 (© OECD/IEA, Explore 2014).

Increasing greenhouse gas concentration in the atmosphere could have potentially significant impacts on world climate. Earth's surface temperature is programmed to rise by 0.6 degrees during the twenty-first century, with the current concentration level of greenhouse gas in the atmosphere, and without taking into account the additional accumulation from "anthropogenic forcing".

Greenhouse gas concentration on the planet shall thus be much higher by the end of the century. So far the alerts from the IPCC scientists have not turned into global and coordinated action plans. National policies often continue to privilege economic competitiveness. The increase of greenhouse gas concentration thus seems to be a historical determinism, almost ineluctable, and with it climate change. The world will not remain as we know it. It is constrained by historical continuities that it cannot manage to control. Humanity watches powerlessly as the world it lives in runs out of resources and diversity.

References

Cisco (2011) The internet of things. http://www.iotsworldcongress.com/documents/4643185/3e968a44-2d12-4b73-9691-17ec508ff67b

Das G (2000) Le réveil de l'Inde. Editions Buchet Chastel, Paris

Durand B (2007) Energie & Environnement, Les risques et les enjeux d'une crise annoncée. Collection Grenoble Sciences, EDP, Paris

EIA (2015) Energy Information Agency. http://www.eia.gov/forecasts/steo/report/global_oil.cfm

Exxon Mobil (2016) The outlook for energy: a view to 2040. http://corporate.exxonmobil.com/en/energy/energy-outlook

FierceWireless (2015) http://www.fiercewireless.com/story/ericsson-backs-away-expectation-50b-connected-devices-2020-now-sees-26b/2015-06-03

Financial Times (2014) http://ftalphaville.ft.com/2014/09/09/1960891/the-dark-side-of-data-centres/

Geohive (2014) http://www.geohive.com/earth/his_history1.aspx

GIEC/IPCC (2007) Rapport de synthèse. http://www.ipcc.ch/pdf/assessment-report/ar4/syr/ar4_syr_fr.pdf

Global Warming (2014) http://globalwarming-arclein.blogspot.com/2013/06/pending-chinese-population-contraction.html

© OECD/IEA, Buildings (2013) Transition to sustainable buildings. IEA Publishing. License: www.iea.org/t&c. As modified by V. Petit. http://www.iea.org/publications/freepublications/publication/transition-to-sustainable-buildings.html

© OECD/IEA, CO_2 (2013) CO_2 emissions from fuel combustion highlights 2013. IEA Publishing. License: www.iea.org/t&c. As modified by V. Petit. http://www.iea.org/publications/freepublications/publication/co2emissionsfromfuelcombustionhighlights2013.pdf

© OECD/IEA, Explore (2014) Explore. IEA Publishing. License: www.iea.org/t&c. As modified by V. Petit. http://www.iea.org/etp/explore/

© OECD/IEA, Non-OECD (2012) Balance of non OECD countries. IEA Publishing. License: www.iea.org/t&c. As modified by V. Petit. http://www.iea.org/media/training/presentations/statisticsmarch/balancesofnonoecdcountries.pdf

© OECD/IEA, OECD (2012) Balance of OECD countries. IEA Publishing. License: www.iea.org/t&c. As modified by V. Petit. http://www.iea.org/publications/freepublications/

© OECD/IEA, Statistics (2015) Statistics. IEA Publishing. License: www.iea.org/t&c. As modified by V. Petit. http://www.iea.org/Sankey/#?c=World&s=Final consumption

© OECD/IEA, Transport (2009) Transport 2009. IEA Publishing. License: www.iea.org/t&c. As modified by V. Petit. http://www.iea.org/publications/freepublications/publication/transport2009.pdf

© OECD/IEA, WEO (2012) World Energy Outlook. IEA Publishing. License: www.iea.org/t&c. As modified by V. Petit. http://www.worldenergyoutlook.org/publications/weo-2012/

Koomey J (2014) http://www.koomey.com/research.html

Larousse (2014) http://www.larousse.fr/encyclopedie/divers/transition_d%C3%A9mographique/187352

Le Monde diplomatique (2010) http://www.monde-diplomatique.fr/2010/04/GOLUB/19008

Lewin M (1985) La formation du système soviétique. Editions Gallimard, Paris

Milankovitch M (1920) Théorie mathématique des phénomènes thermiques produits par la radiation solaire. Gauthier-Villars, Paris

National Academies (2009) http://www.nationalacademies.org/includes/G8+5energy-climate09.pdf

OMM (2014) Organisation Météorologique Mondiale. http://www.wmo.int/pages/mediacentre/press_releases/pr_1002_fr.html

Planetoscope (2014) http://www.planetoscope.com/natalite/5-.html

Shell (2016) New Lens Scenarios. http://www.shell.com/promos/english/_jcr_content.stream/1448477051486/08032d761ef7d81a4d3b1b6df8620c1e9a64e564a9548e1f2db02e575b00b765/scenarios-newdoc-english.pdf?

Todd E, Le Bras H (2013) Le mystère français. Editions le Seuil, Paris

World Nuclear Association (2014) http://www.world-nuclear.org/info/Current-and-Future-Generation/World-Energy-Needs-and-Nuclear-Power/

The Energy Industry: Running at Full Speed

3

The world faces new challenges in energy production. The sharp increase in demand linked to population growth and new economies leads to higher demands on resources and production means. The "energy machine" is running at full speed and ensuring a sustainable supply of the rising demand is the day-to-day challenge of the energy industry.

3.1 Oil as the Main Primary Energy Source

At the heart of those issues is oil. Representing over one third of total primary energy consumption, it is the primary energy source.

3.1.1 Towards a Further Concentration of Oil Production

World annual oil production in 2013 was around 4100 Mtoe (BP 2014) (Fig. 3.1).

After the two oil crises in the 1970s (1973 and 1979), the world diversified its sources of oil production. The Middle East represents today 32% of total oil production, Eurasia 15%, and North America 19%. A number of smaller producers coexist with these giants, such as Europe (5%), Africa (10%), South America (9%) and China (5%). North America has always been a large producer but has successfully managed to maintain its share thanks to the development of unconventional oil in the United States and Canada (Fig. 3.2).

Global oil production has increased on average by 1% per year in the last 10 years. This increase hides several disparities. Output from the United States and Canada grew by more than 3% on average, their growth driven by the development of unconventional oil. That from Eurasia increased by an average of 2.4%, notably in former Soviet Union countries. Middle East production grew by an average of 1.6% per year. This region alone represents 50% of the total world production increase. All other sources of oil production which had developed from

© Springer International Publishing AG 2017
V. Petit, *The Energy Transition*, DOI 10.1007/978-3-319-50292-2_3

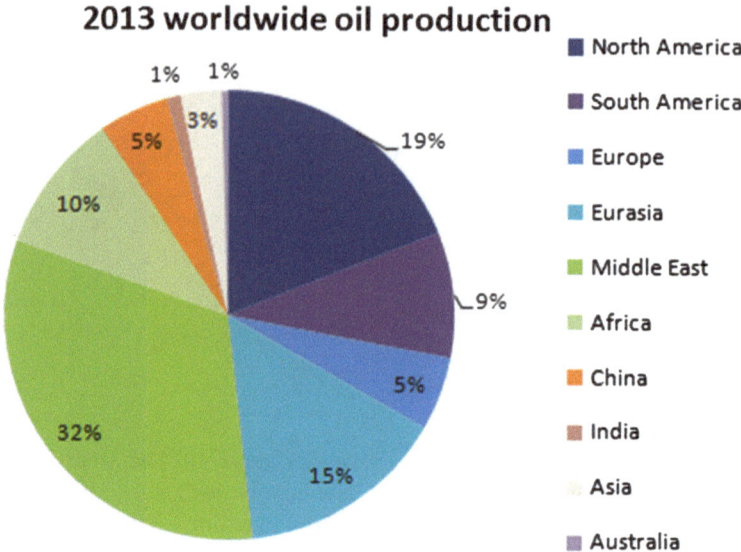

Fig. 3.1 Worldwide oil production (BP 2014)

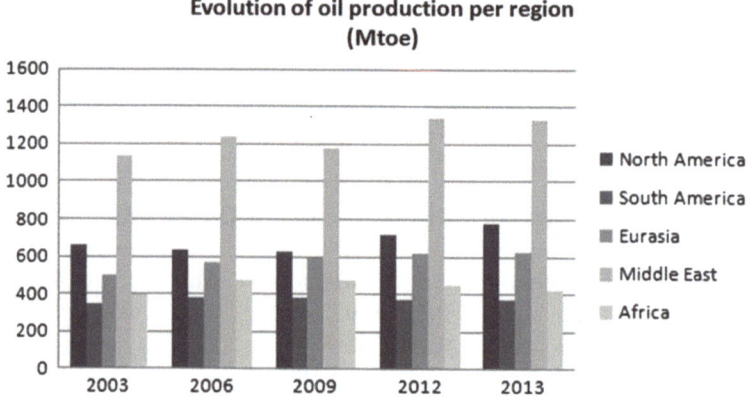

Fig. 3.2 Worldwide oil production per region (BP 2014)

the 1970s have either stagnated or decreased in the last 10 years. While South America's output grew, it was by no more than 0.9% on average per year. Africa also became a bigger producer, but by less than 0.5% per year on average. European production has steadily decreased by an average of 0.5% per year. This means that oil production has concentrated among a few players in the last 10 years. The combined share of North America, Middle East and Eurasia rose from 61% in 2003 to above 66% today (Fig. 3.3).

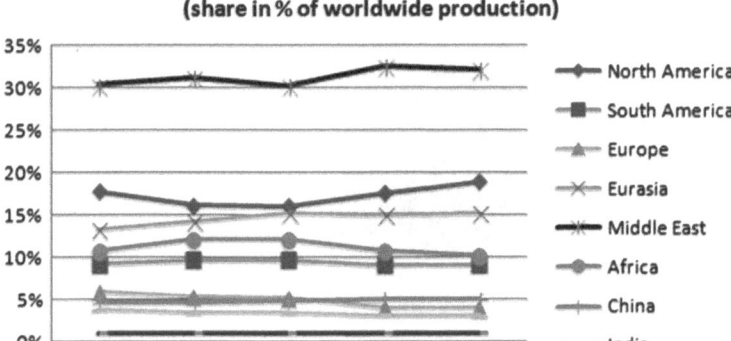

Fig. 3.3 Share of oil production per region (BP 2014)

So, despite oil price being fairly high in the last 10 years and growing demand from new economies, oil production has concentrated around a few actors, a rather counterintuitive development partially caused by unconventional oil developments as well as oil price speculation.

3.1.2 Shift of Oil Consumption to Asia

North America remains the top region for oil consumption with more than 25% of the total world consumption. Europe is second with 17%. China and the rest of Asia represent a third of total consumption, with demand rising fast (Fig. 3.4). The North American continent is relatively autonomous when it comes to production, while Europe and Asia share the resources from Eurasia and from the Middle East (and, to a lesser extent, Asia itself and Africa).

In the last 10 years, all mature economies have reduced their oil consumption—North America by an average 0.8% per year, Europe by 0.3% per year and Japan by more than 4% per year. New economies have however increased their consumption in a spectacular manner. China's consumption increased by more than 6.5% per year, India by 4.1% per year, South America's and Africa's by more than 3% per year (Fig. 3.5).

Actually, the share of North America and Europe in the world's oil consumption reduced by more than nine points in 10 years (from 50 to 41% of total consumption), while China gained five points and South America and Middle East two points each. The other regions remained stable. Asia hides many disparities: Japan represents one third of the region and its consumption dropped by more than 4% per year while demand from the rest of Asia grew by more than 2% on average. Global economic growth drives oil consumption, but mostly it is the development of new economies which led to the global increase in oil consumption. In the end, the

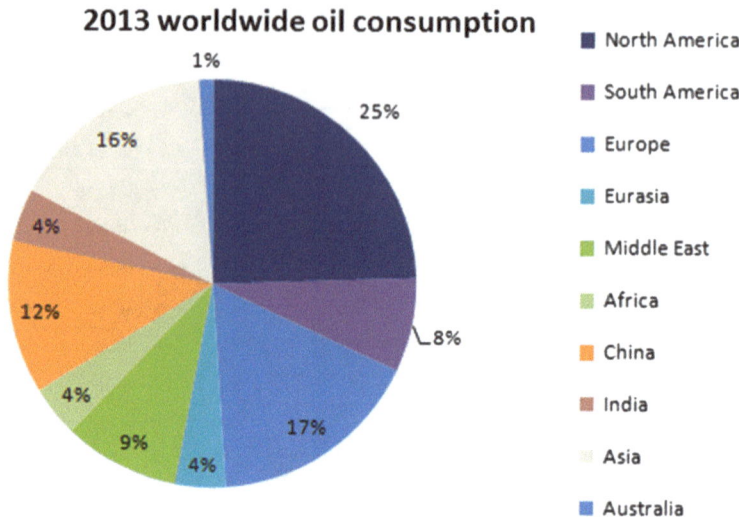

Fig. 3.4 Worldwide oil consumption (BP 2014; © OECD/IEA, WEO 2012)

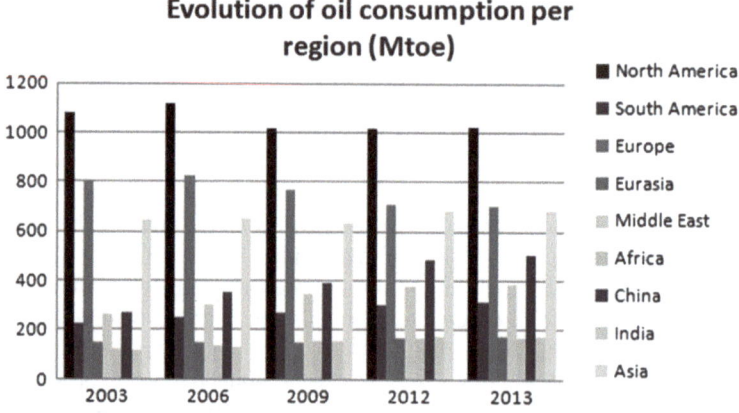

Fig. 3.5 Worldwide oil consumption per region (BP 2014; © OECD/IEA, WEO 2012)

economic transition confirms that oil consumption is shifting from North America and Europe to Asia.

3.1.3 Oil Geopolitics

Today, 57% of the oil consumed in the world is exchanged across regions.

Europe and Asia (including Japan, China and India) are the two main importing regions. The needs of Asia are primarily covered by the Middle East, which accounts for 53% of demand; 75% of oil exports from the Middle East are directed to Asia. In Europe, 41% of the needs are covered by imports from Eurasia. The rest comes from the Middle East (17%) and Africa (23%). A third of the oil that North America imports comes from the Middle East. Now, this volume represents only 10% of the total oil consumption in North America, which shows the limited dependency of North America on oil imports in general, and on the Middle East in particular (BP 2009, 2014).

The geopolitics of oil are thus made up of a variety of situations. The American continent is relatively autonomous, the Middle East region is essentially linked to Asia, and Eurasia (in particular Russia) remains strongly tied to Europe. If North and South America continue to be autonomous in oil and the interdependency between Asia and the Middle East endures, the relationship between Russia and Europe could change—Russia could turn more towards Asia, which energy needs are growing considerably (Table 3.1).

3.1.4 Fast-Forward in the Next 20 Years

There are several scenarios for how oil production might evolve in the next 20 years (© OECD/IEA, WEO 2012). The first, in which status quo is maintained, estimates that global oil production will grow from 87 million barrels daily to above 108 million barrels by 2035. A second scenario, in which a number of measures will be adopted to limit global warming to less than two degrees above the pre-industrial level, envisages a reduction of total production to around 79 million barrels per day. Finally, an intermediate scenario ("New Policy" scenario) sees a slight increase of production to a bit less than 100 million barrels per day. This scenario is presented here.

According to the International Energy Agency (2012), oil production should increase on average 0.5% per year. This forecast corresponds to other forecasts such as Exxon Mobil (2016) with 0.7% growth in average per year, Shell (2016) (0.7% for the "Mountains" scenario), or Statoil (2016) with similar growth. Output in the Middle East should grow twice faster and the region's share of oil production should thus increase from 32 to 36%. North and South America should see their production increase by an average of 1%, and consequently their share, in particular North America's, should as well increase in the overall production mix. The growth of production in Eurasia should remain very weak, mainly due to the lack of investment in the region. Without any new significant investment in the next 20 years, only 10 years of available production would remain available to Eurasia in 2035 (Fig. 3.6).

The world's oil consumption should increase as well by an average of 0.5% per year, with strong regional disparities. Large industrial regions such as North America and Europe should lower their energy consumption by an average of 1% per year. This would take their share of global energy consumption down from 41 to 30%. The rest of the world should consume more energy, notably China (2.2% per year) and India (3.4% per year). In Asia, Japan should consume less while the rest of Asia demands more energy. By 2035, the share of Asia (including China and

Table 3.1 Worldwide oil exchanges (BP 2009, 2014)

Oil (Mtoe) Transit energy from	Transit energy to										
	North America	South America	Europe	Russia/Eurasia	Middle East	Africa	China	India	Asia	Australia	Total exports
North America	×	56	46	0	4	7	10	7	16	0	146
South America	85	×	18	0	0	0	31	32	18	0	184
Europe	36	9	×	5	12	29	1	1	23	0	116
Russia/Eurasia	25	1	295	×	13	2	43	2	41	2	424
Middle East	107	7	103	0	×	17	154	125	444	8	965
Africa	48	19	147	0	1	×	65	32	19	7	338
China	0	5	1	1	1	1	×	1	22	0	32
India	3	5	8	0	17	9	×	×	17	0	60
Asia	7	2	6	0	3	5	34	5	×	34	96
Australia	0	1	0	0	0	0	3	0	10	×	14
Total imports	311	105	624	6	51	70	342	205	610	51	2375
Consumption	1024	312	711	167	385	171	507	175	686	47	4185

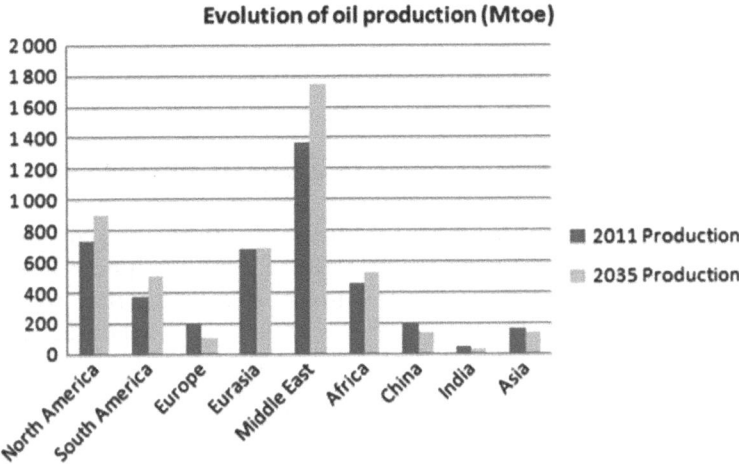

Fig. 3.6 Evolution of oil production (© OECD/IEA, WEO 2012)

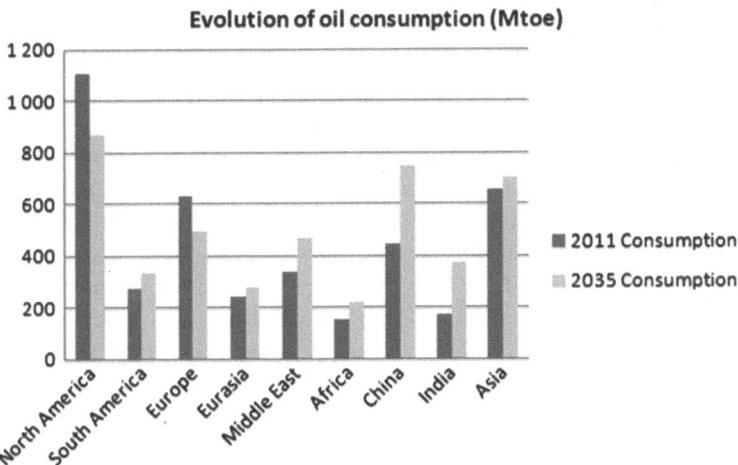

Fig. 3.7 Evolution of oil consumption (© OECD/IEA, WEO 2012)

India) will represent 40% of world oil consumption, versus 30% today. The shift from West to East is obvious. Countries from the Middle East and Africa will continue to see their consumption rise, but not in the same proportion. In Africa, the rise of oil consumption will essentially be linked to population growth rather than the economic transition expected to start later in this scenario (Fig. 3.7).

The oil production and consumption pattern in the last 10 years is thus expected to continue for the next 20 years. The economic transition in Asia will materialize and the shift of consumption towards Asia will amplify. Consequently, oil production in the Middle East will continue to increase, as well as in Russia if China and

Russia develop stronger ties in the coming years. Finally, the development of unconventional oil in the American continent shall lead to an increase in production, confirming the relative energy independence of the region.

3.1.5 "Peak Oil": Towards a Possible System Breakdown?

There are many who question if there will be enough reserves to actually fuel economic growth. They refer to a possible date where production will inexorably start to decline as "peak oil". "Peak oil" is first of all technical data, which can be measured for every field. There is indeed a point in time in field extraction where production maxes out. Progressively, as resources diminish, it becomes more difficult to extract the remaining oil and daily production declines. Conditions of pressure and access to resources make it indeed more complicated to collect the remaining oil after the "easy" oil has been extracted. The world's "peak oil" thus adds up all production "peak oils" from all fields in the world. Evaluating "peak oil" is very complicated due to the relative lack of knowledge of each field's geology. The exact volume of reserves is unknown, only estimated, and the actual quantity of oil that can be extracted is even more difficult to predict because of the geological particularities which can limit or facilitate field extraction.

This technical definition of "peak oil" only accounts for "proven" reserves. "Probable" resources then need to be added to gather the overall "retrievable" resources. Resources are estimated to be "retrievable" with current state-of-the-art technologies. They can consist of improvements over existing fields or operation of new fields. Many uncertainties remain over the actual capacity to recover totally these reserves, which is why they are not classified as "proven". Everything has been said and written on "retrievable" resources. Oil exploration intensified following the two oil crises of the 1970s. Large oil companies and governments tried to diversify their procurement to limit their dependency on the Organization of the Petroleum Exporting Countries (OPEC) countries. Many limits and roadblocks were overcome: geographically, exploration reached new frontiers (Latin America, Africa, the Arctic, deep offshore) and technically, the development of unconventional oil (extra heavy crude, shale oil, etc.) radically modified the oil resources landscape. Actually, the increase in reserves is mainly due to significant efforts realized in these domains. Since 1970, one trillion barrels have thus been discovered, representing 40 years of global oil consumption. However, annual oil consumption (around 36 billion barrels a year) today exceeds new discoveries. Therefore, the volume of reserves is diminishing. In addition, 70% of the increase in "proven" reserves (© OECD/IEA 2014) has come from the reevaluation of existing fields.

There are today 1.7 trillion barrels of "proven" reserves (BP 2014), which corresponds to 238 billion tons. There are also about 2.6 trillion barrels of conventional "retrievable" reserves (Furfari 2007), or 364 billion tons. Unconventional resources (asphaltic crude, extra heavy crude or shale oil) represent 3.1 trillion barrels of "retrievable" resources (or 435 billion tons), which can also be added to

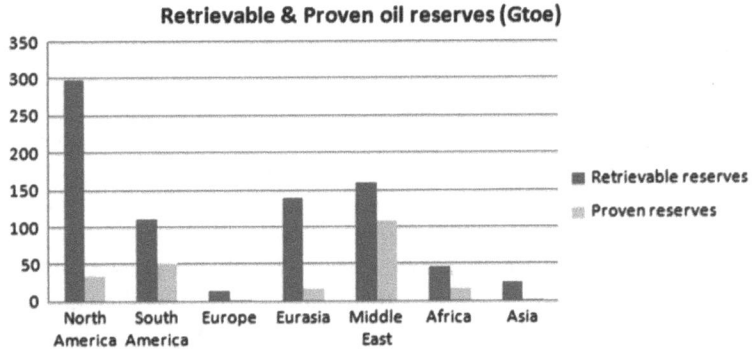

Fig. 3.8 Oil reserves (BP 2014; Furfari 2007; © OECD/IEA, WEO 2012)

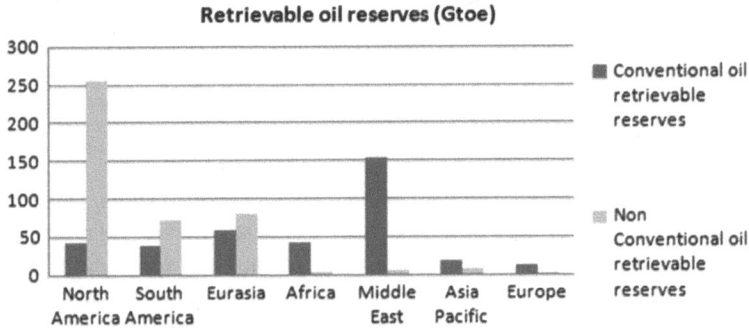

Fig. 3.9 Retrievable oil reserves (BP 2014; Furfari 2007; © OECD/IEA, WEO 2012)

the previous figure, leading to an overall 5.7 trillion barrels of overall "retrievable" oil reserves (© OECD/IEA, WEO 2012), or 795 billion tons (Figs. 3.8 and 3.9).

World consumption today is up to 100 million barrels per day. Assuming zero increase in consumption, current oil reserves should correspond to around 150 years of production. Reserves are often measured using only "proven" reserves, which amounts up to 1.7 trillion barrels (BP 2014), or 50 years of production at the current consumption rate. However, the figure presents a wrong perspective of the reality as it implies "retrievable" reserves cannot actually be recovered and does not account for unconventional oil reserves.

To sum up, today, there are in reserve 50 years of conventional oil production at the current rate, likely twice more when conventional "retrievable" reserves are taken into account, and finally around 150 years of production when non-conventional oil reserves are included.

The discovery and exploitation of non-conventional oil in North America disrupted the balance of power in oil that had been in place since World War II. Henceforth, the core of worldwide oil reserves is situated in North America. The huge gap between identified reserves in North America and elsewhere also suggests that exploration may not have been conducted completely in other regions of the world, where additional reserves could be identified in the coming years.

While new "retrievable" oil reserves have been identified, their discovery happens at a slower rate than consumption. Still, the total amount of "proven" reserves increases year after year. There will therefore not be any catastrophic oil shock in the coming years. Be that as it may, a number of events could possibly impact and raise tension in the overall state of reserves and worldwide oil production. Geopolitical events could indeed influence access to resources and therefore prices, available resources, and exchanges across countries. The speculation in oil prices as well as the reactive measures that could be taken by the various market players could impact the investment capacities of oil companies; investment in new oil fields is the only way to replace old fields which have reached end of life. Beyond oil price, political decisions to diversify the mix of oil imports in high-consumption countries could also influence the overall flows of oil exchanges and therefore the access to investments from specific regions of the world. This could in turn have an impact on overall production capacity. The "peak oil" might therefore well happen, but not likely because of a lack of "retrievable" resources. More likely, it might happen because of a sustained imbalance of investments compared to the evolution of demand.

In summary, the "peak oil" is technically interesting but often dealt with too simplistically. The question of "peak oil" actually relates to the balance between consumption, production and the discovery of new reserves. Today, reserves continue to grow, albeit at a slower pace than that of consumption. This pace could pick up if more exploration investments were triggered. Production adapts to consumption forecasts, and the necessary investments are only triggered if they are considered profitable with regards to oil price. The "peak oil" could thus occur from "proven" reserves progressively drying out simply because of a lack of investments. However, it will almost certainly not occur from exhaustion of "retrievable" resources. The "after oil" period is ahead of us and what will happen depends on a multitude of geopolitical, economical and technical factors. The question is a delicate one as the combination of these factors remains uncertain. In the end, we could quote John Mitchell (Furfari 2007), an expert in oil economics matters: "oil reserves are unknown, cannot be known, and are not important".

3.1.6 Oil Price Swings

According to the Goldman Sachs Commodity Index (Furfari 2007), oil is the top resource exchanged in the world, accounting for 43% of the total volume of exchanged resources. In comparison, the most important non-energy resource shared is livestock, with only 3% of the total.

OECD economies have developed for a century industrial flagships (the "Majors") to explore, produce and distribute oil to their nation of origin and therefore sustain economic development. These "Majors" have long dominated the oil trade as they were the only companies capable of mastering the technicalities of oil field development as well as having the financial capacity to execute such projects. This domination largely favored oil-consuming nations, and was rather detrimental to the countries where the oil was produced. Wealth distribution was imbalanced in favor of OECD economies. This was the time of very cheap energy, which ended in the 1970s. Countries where the oil was produced regrouped in OPEC and took back control of their national reserves. They created their own companies or nationalized the subsidiaries of the "Majors" operating on their soil, in effect taking back control of the oil economic rent (WTRG 2013). Today, only 20% of the yearly 600 billion dollars of investments in oil production and distribution are from the "Majors". The vast majority of the investments are controlled by national companies, the new decision centers.

Ricardo's rent theory (Economic Theories 2008) explains that, to meet a specific level of demand, the most economical fields are operated first, then the more expensive ones. The price of oil is fixed by the marginal cost of production of the field with the highest cost that still needs to be operated to produce the expected demand. All other fields benefit from a rent as their marginal cost of production is lower. When OPEC took control of its fields, it changed the way prices were determined. As a result, the general theory from Ricardo proved to not be applying anymore. Oil was produced in the Middle East because its fields were the cheapest to operate. OPEC decided to change the pricing system by setting production quotas. As a result, prices went up and the Middle East region benefited from a higher rent. When the OPEC countries made this change, the "Majors" started to look at diversifying their resources in order to limit their geopolitical dependency. Higher prices led to fields with higher marginal costs and new fields started to be developed in South America, in the North Sea and in Africa.

A large share of oil production is sold through long-term contracts (in general, of 1–2 years in duration); the rest is sold on the "spot" market. The size of the "spot" market is not exactly known (Financial Times 2013), and has been estimated at between 10% (Furfari 2007) and 40% (Desjardins 2007) of the total of the volumes physically exchanged (around 100 Mb/d). A number of financial tools have also been developed in the oil trade. "Future contracts" allow the various players to reduce their risk or speculate with regards to the oil price evolution. The development of these financial tools has attracted a number of new players in recent years. Those new entrants are essentially hedge funds, which play a role in accentuating "spot" market volatility. Price volatility can reach up to 40%, much higher than the stock market (20%) or the obligations' market (12%) (Furfari 2007). The price variations on the "spot" market as well as the share of the volumes that each company (national or international) trades on the market versus the long-term contracts that they enter into influence the revenues of these companies, and therefore their investment capacities (and often, in the case of national companies, the country's economy). Additionally, the volatility of prices makes the calculation of the profitability of an investment more complicated, especially when this investment amounts to several billion dollars and lasts several years (typical payback period for such investment is 6–8 years). However, despite their volatility, oil prices can be explained and, to a certain extent, forecasted.

3.1.6.1 The Logic in Oil Price Evolution

The oil market is a global commodity market. It is thus influenced by a variety of factors. The graph below models the various factors that can influence oil price variations.

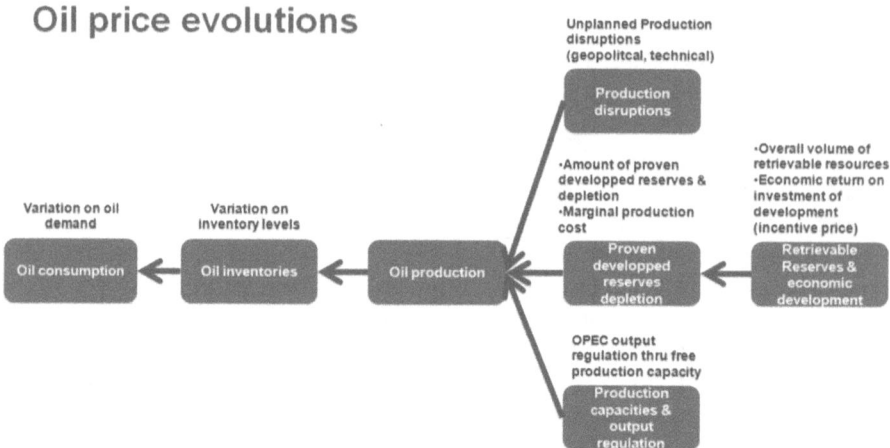

One of these elements is, of course, demand. Demand is affected by a multiplicity of factors, both short-term and mid-term. Demand is first a result of short-term economic activity, which drives more or less transportation or industrial needs. Weather and seasons can also be an important factor—oil is used for heating (oil represents 20% of district heating energy consumption). Mid-term demand patterns are more related to the economic stability and development of nations. The transportation sector accounts for 54% of all oil demand and is thus critical in predicting oil consumption evolutions. Industrial development comes second, accounting for over 31% of total oil consumption.

Another element which influences the price at which oil is purchased from producing countries is the level of oil inventories. They generally range around 50 days of consumption but have evolved in the recent years. A structural increase in inventories pushes prices down while relatively low inventories tend to pull prices up.

Also, oil production varies significantly due to a variety of factors. Production disruptions are frequent, sometimes due to geopolitical tensions. Wars in the Persian Gulf in the 1980s and 1990s, in Iraq in 2003, and in Libya in 2011 have disrupted global oil production. Tensions in Africa (Nigeria, Angola, etc.) regularly lead to production disruptions. Geopolitical tensions can also take the form of economic embargos, such as the recently lifted one on Iran, which led to reduced output globally. Disruptions can also be the result of technical issues. Most production facilities experience such issues and they may have a strong impact on short-term price movements.

In addition, oil price is partially controlled by how the OPEC countries regulate their oil outputs. Although not widely respected, the system of production quotas shared among the OPEC members influences overall production output. Saudi Arabia, being the first producer in the world, as well as the cheapest source, has been in recent years the main factor of production adjustment. Regulating production up or down has influenced significantly the oil price level. While regulating down production helps increase prices, this makes it difficult for OPEC countries when it comes to regulating down prices, due to a reduced free production capacity.

Finally, oil production obviously depends on the production capacity and the depletion of reserves. Proven developed reserves correspond to fields already in operation. The price to produce one more barrel is called the marginal cost of production. The marginal cost of the last barrel to be produced to meet demand is thus a key indicator for setting oil price. Below the marginal cost of production, producers start to shut down fields as it becomes uneconomical to operate them, and the remaining fields progressively get depleted. When demand continues to grow, it becomes essential at a certain point in time to restart fields or to develop new reserves (retrievable) to balance offer and demand. The price at which the fields can be restarted or the retrievable reserves be developed and made productive is, however, much higher than the marginal cost, as it takes into account the return on investment for the operator. The breakeven price for the operator (or incentive price) in an uncertain market is thus a key factor to restarting production in order to meet demand. Restarting a field is less costly than developing a new field (BNP 2015). The typical payback for investments in new fields averages 8 years, with the notable exception of shale oil in the United States, for which payback is only 2 years. An accurate and sustainable oil price level is extremely important to trigger the massive capital expenditure investments that the development of a new field requires.

To sum up, oil price depends upon a multiplicity of factors. Some will affect short-term oil price variations, such as:

- oil inventory levels
- short-term unplanned production disruptions (technical, geopolitical)
- spare capacities of OPEC countries and their capability to regulate output
- marginal cost of production

while a few other factors will influence oil price variations in the mid-term:

- consumption demand evolution and its responsiveness to oil price evolutions
- the economic soundness to develop new production capacities to renew depleted reserves

These two oil price variations cycles obey different rules, and must thus be looked at separately.

In conclusion, the difficulty in forecasting oil price relates to the fact that despite the market being a global and well-organized one, the data needed to accurately forecast the behavior of the various market players is often missing. The amount of proven developed reserves is generally not transparently communicated. Reserves are often estimated but not accurately known. The marginal costs as well as the breakeven costs are also not accurately known and shared. This lack of transparency generally leads to wrong interpretations. Production disruptions, especially when they are of a technical nature, are difficult to predict, although they have an important impact on the final price. Forecasting oil prices is thus an imperfect exercise.

3.1.6.2 Short-Term Oil Prices: The Example of the 2014/2015 Price Drop

In 1 year, oil prices dropped by over 50%. This significant change of paradigm for oil companies is the result of short-term factors which combined together at the same time. It is a good example for understanding how oil prices can vary in short-term cycles.

As already explained, oil price cycles in the short term are influenced by a number of factors. Oil prices are first determined by the marginal production cost of the last barrel of oil needed to meet demand. All the barrels of oil which can be produced at a lower cost are first supplied, and the producer's profit is based on the last barrel's marginal cost to meet the demand. The marginal cost of oil is thus theoretically key to oil price. As consumption evolves upwards or downwards due to short-term economic cycles, the cost of production of the last barrel of oil can evolve significantly. The graph below shows the various marginal costs of production of oil per region and type of production (Knoema 2014) (Fig. 3.10).

Shale oil can be further detailed as the market is not homogenous. Marginal costs can be as low as 30 USD/bl and may go up to 80 USD/bl (Telegraph 2014). The particularity of the United States' shale oil market is it is shared by a large number of small suppliers. Market consolidation has yet to happen, and with it a drop of operation costs, although the current shale oil crisis in the United States has not

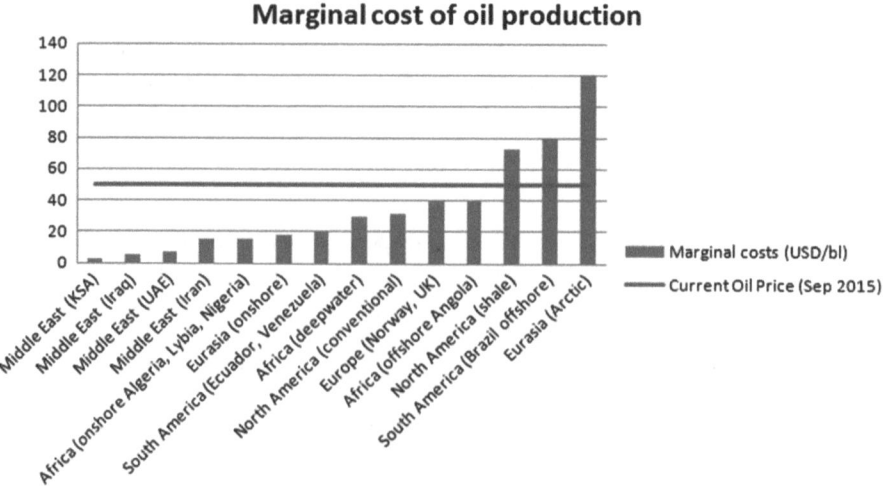

Fig. 3.10 Marginal cost of oil production (Knoema 2014)

shown any indication of an upcoming consolidation (World Economic Forum 2015).

It is clear that most oil-producing countries continue to profit with a current oil price of 50 USD/bl (end of 2015), even though profits have been halved during 2015. The oil market can thus still continue to operate at such price levels, at least in the short term. The International Energy Agency (2015) estimated that the volume of oil production that could be shut down due to non-profitable operations could actually be as low as 0.4 Mb/d at 50 USD/bl, and close to 1 Mb/d if the price goes down to 45 USD/bl.

In the case of the 2014/2015 oil price drop, the marginal cost of production thus has turned out to have a limited impact on the overall production output. This limited volume of missing production could in addition, in the short term, easily be supplied by other sources, notably the Middle East, with rising production levels in Iran, Iraq or Libya. Actually, the sudden increase of production from "cheap" sources from the Middle East has had in the last year a considerable impact on the oil price level, by offering a cheaper alternative to the market. In the last 3 years (2012–2015), oil production has indeed increased yearly by around 2 Mbl/d, much faster than consumption, which increased only by 1 Mbl/d. This is the imbalance between offer and demand which pushed prices down.

During this period, the "free" capacity of OPEC countries has also remained stable at around 2 Mbl/d (EIA 2015), thereby not pushing prices back up. This "free capacity" represents 2 years of worldwide increase in oil demand (Fig. 3.11).

Another element which pressed oil prices down is the volume of inventories (© OECD/IEA 2015; Macrotrends 2015). With the variation of supply and demand, inventories were being built or emptied accordingly. According to the International Energy Agency (2015), the level of inventories in OECD countries grew by 715 Mbl between 2014 and 2015, to reach a record high of 2923 Mbl, and it should keep growing in 2016 as there is no indication that storage capacity limits have been reached (Fig. 3.12).

The high level of inventory has a structural impact on oil prices as demand may become more selective. Figure 3.13 maps both the level of inventory (© OECD/IEA 2015) and oil price (Macrotrends 2015). The correlation between the two is obvious. As the level of inventories increased, a sign of oversupply (or lower demand), the price followed a downward curve.

Finally, unplanned production disruptions also affect dramatically the oil price, as they reduce output in an unexpected manner, leading prices up. This "missing" volume by nature cannot be forecasted but only estimated. Data from the United States Energy Information Agency (2015) shows that this volume is greatly influenced by conflicts and geopolitical tensions. There has been a clear tendency towards an increase of this volume year after year. The overall volume of unplanned production disruptions grew from 1.5 Mbl/d in 2012 to close to 3.5 Mbl/d in 2015. The volume of unplanned production disruptions in OPEC countries jumped from 1 Mbl/d to close to 3 Mbl/d in the same period (EIA 2015). This increase, however, did not impact the overall oil price drop in a significant manner in the period; rather it created the conditions for increased volatility in the market (Fig. 3.14).

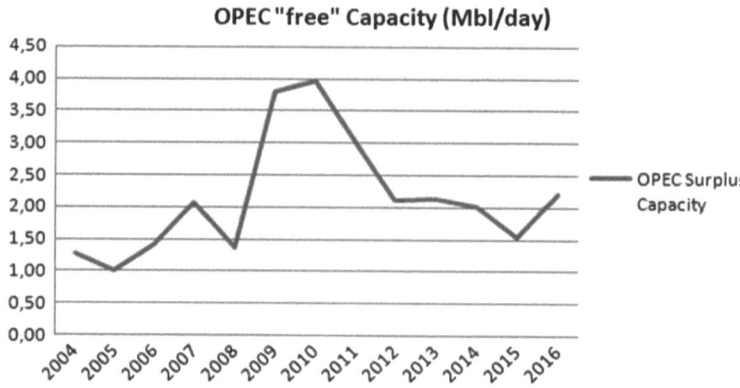

Fig. 3.11 OPEC free capacity (EIA 2015)

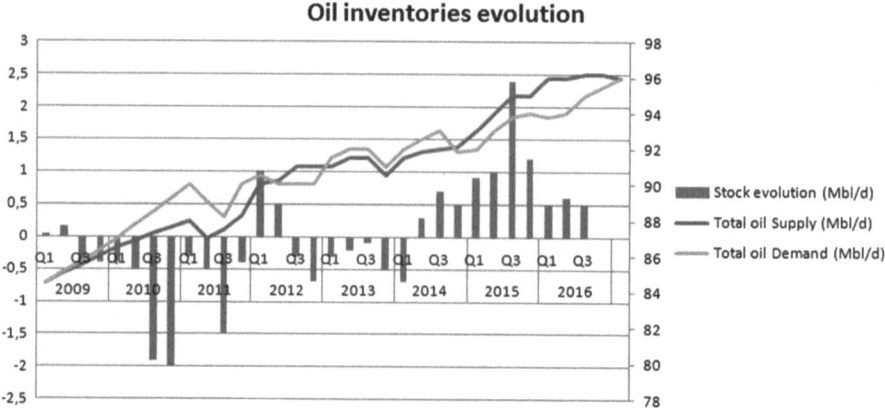

Fig. 3.12 Oil inventories (© OECD/IEA 2015; Macrotrends 2015)

In conclusion, the surge of oil production from the Middle East and from the United States (shale oil) contributed to fill up the inventories at a record high level, pushing prices down. Due to the low marginal cost of production of most oil producers (including shale oil companies in the United States), the level of production did not adjust significantly with the price (and the marginal cost of production played a little role in the oil price evolution since it was much lower than the actual price of 2014). In addition, the stable "free" capacity of OPEC countries contributed to maintain prices at a low level since the market apparently managed to adjust to this new price level. Traditionally, Saudi Arabia (and the other OPEC countries) reacted to such evolutions by reducing their production output. They decided to not do so this time, leaving prices at around 50 USD/bl.

If oil consumption continues to grow by 1 Mbl/d on average per year while overall oil production decreases by 0.5–1 Mbl/d due to the price of one barrel

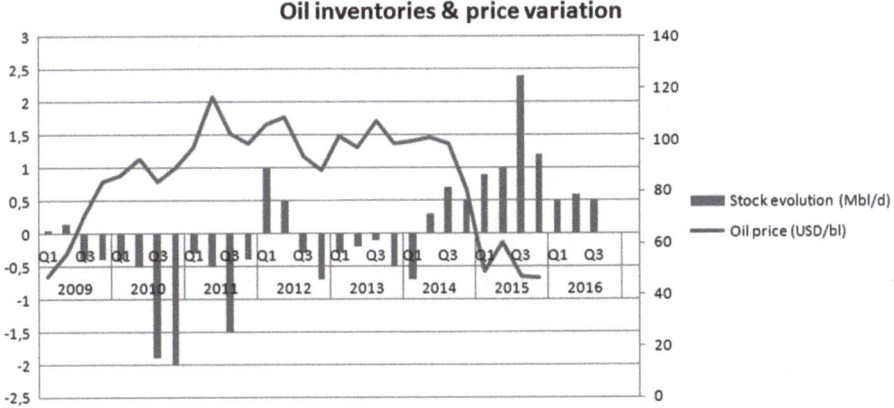

Fig. 3.13 Oil inventories and prices (© OECD/IEA 2015; Macrotrends 2015)

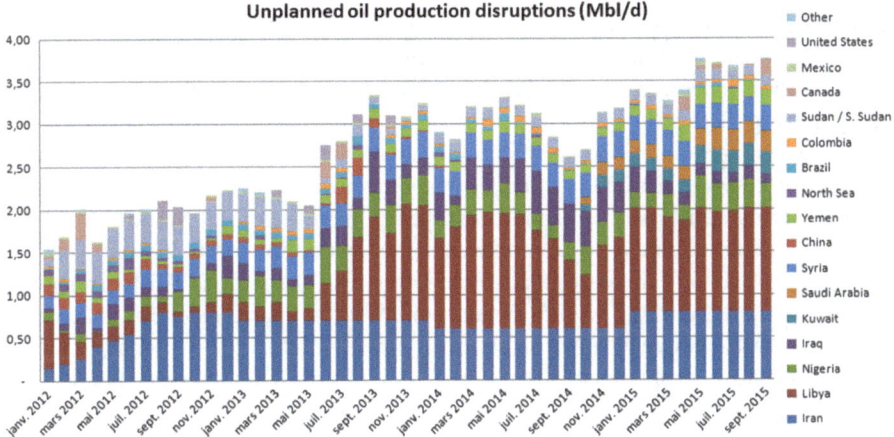

Fig. 3.14 Unplanned oil production disruptions (EIA 2015)

staying around 50 USD, there will be a deficit of production of around 1.5–2 Mbl/d overall; this deficit will increase year after year. The "free" capacity of OPEC countries can sustain the increase in demand in the short term, and it will take a few years to bridge the gap between production capacity and demand. During this time, prices are likely to stay low. The increase of unplanned production disruptions could also lead to increased volatility in the market. Three years ago, the "free" capacity of OPEC was indeed able to compensate for unplanned oil production disruptions. This is not true anymore. The volatility in prices is thus expected to increase significantly during this period.

In the long run, additional production capacities are required to meet the increased demand. These investments which in practice have payback periods of

6–8 years (with the exception of shale oil in the United States) will require per-barrel price to be sufficiently attractive. The oil price is indeed not only the result of short-term balance between offer and demand; it is also an indicator of the necessary production capacity increase foreseen by the market.

3.1.6.3 The Mid-Term Oil Price: The Need to Trigger Investments Must Eventually Pull Prices Up

The first factor which influences oil prices in the mid term is, of course, demand. The evolution of oil consumption in the coming decades will greatly determine the price level at which oil is supplied. Indeed, the oil price determines the profitability (or non-profitability) of new oil fields.

The International Energy Agency (2012) has come up with three scenarios to forecast the oil consumption evolution in the next two decades. The "Current Policies" scenario forecasts oil consumption evolution without any change to current policies and trends. It is the high-base scenario. The "New Policies" scenario forecasts oil consumption evolution taking into account the policy changes in progress which have a high chance to actually happen. It is the realistic scenario. The "450" scenario forecasts oil consumption evolution as it should be to meet global carbon reduction targets and limit CO_2 concentration in the atmosphere to 450 ppm (Fig. 3.15).

The "Current Policies" scenario plans for an increase of oil consumption of 0.9% per year on average till 2035, which corresponds to an overall increase of 25% in absolute terms. This also corresponds to a yearly increase of 0.9 Mbl/d. The "New Policies" scenario plans for an increase of consumption of 0.5% per year on average, or 15% total increase in absolute terms over the period. This corresponds to a yearly increase of 0.5 Mbl/d. Finally, the "450" scenario plans for an average decrease of 0.4% per year, or 9% in absolute terms over the 2010–2035 period. This corresponds to a drop of consumption of 0.3 Mbl/d. The International Energy Agency forecasts are thus quite conservative, considering that the average increase in oil consumption in the last 3 years has exceeded 1 Mbl/d (EIA 2015). The trend definitely looks upwards.

The growth will mostly come from Asia, in particular China and India, as well as the Middle East to a lower extent. This considerable growth (70% in China, 124%

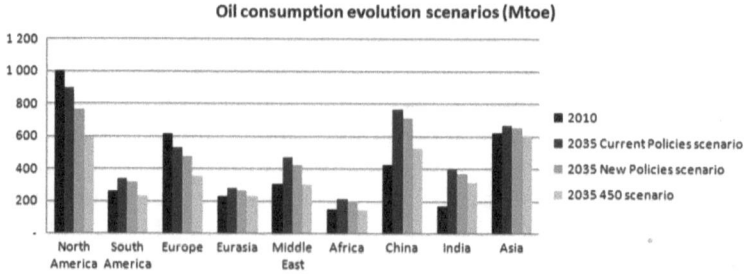

Fig. 3.15 Oil consumption evolutions (© OECD/IEA, WEO 2012)

in India, 40% in the Middle East) will be strongly compensated by the drop in consumption of North America (−24%) and Europe (−23%) over the same period, following the implementation of energy efficiency measures, in particular in the transportation sector. This is why in the end the global oil consumption evolution is expected to be limited. Of course, further disruptions can also happen. They will be described in Chap. 5. They actually correspond to some of the possible solutions to meet the "450" scenario and would lead to a more significant reduction in oil consumption.

In the end, government policies will be instrumental to shaping up a sustainable environment of constraints which will facilitate these transitions and influence the evolution of oil consumption. The current level of oil prices could have a negative impact on the will of governments and nations to actually make a change, as the situation yields an increase in economic growth at no cost. The drop of oil prices at the end of 2014 indeed yielded in 1 year a massive transfer of wealth from exporting regions to importing countries (Table 3.2).

Overall around 930 billion dollars of wealth were transferred. The figure is calculated using the actual flows of energy traded across regions (BP 2014) multiplied by the differential level of price between mid-2014 and mid-2015. It shows that the Middle East and Eurasia lost the most, while Europe and Asia benefitted the most. North America, being both a producer and a consumer, is not impacted at the same level. In the end, this massive amount of wealth will have an impact on high-consumption regions, driving economic growth and possibly discouraging some initiatives related to energy efficiency. This impact will however remain limited in North America. A number of studies (European Central Bank 2004; IMF 2012) showed the complicated relationship between GDP growth and oil prices. Basically, this relationship is non-linear. The impact of a strong increase in oil prices was found to have a strong negative effect on most oil importing countries, while a decrease actually affects only a few countries. The effects of oil price swings also differed greatly between exporting and importing countries, although not symmetrically. In the end, the recent oil price shock is creating favorable conditions for oil importing countries, which needs to be looked at case by case.

Besides consumption evolution, a second factor that can influence prices in the mid-term is production capacity. The "free" capacity of OPEC countries has significantly dropped in the last 30 years (Conca 2015). While it used to be as high as 15 Mbl/d, it went down in the recent years to around 2 Mbl/d. Production capacity is essential to meet demand. Oil demand increases year after year in almost all the International Energy Agency scenarios, except the "450" scenario. With none of the actual policies required to meet the "450" scenario having been put in practice, the likely trend of oil consumption in the coming years is upwards. Production capacity must thus be increased. This is all the more relevant considering that the "free" capacity of OPEC countries is not enough to compensate for the increase in demand, and taking into account the increased number of production disruptions.

Table 3.2 Wealth transfers across regions (BP 2014)

Billion USD Transit energy from	Transit energy to										
	North America	South America	Europe	Russia/ Eurasia	Middle East	Africa	China	India	Asia	Australia	Total
North America	×	22	18	0	2	3	4	3	6	0	−57
South America	33	×	7	0	0	0	12	13	7	0	−72
Europe	14	4	×	2	5	11	0	0	9	0	−46
Russia/Eurasia	10	0	116	×	5	1	17	1	16	1	−167
Middle East	42	3	40	0	×	7	61	49	174	3	−379
Africa	19	7	58	0	0	×	26	13	7	3	−133
China	0	2	0	0	0	0	×	0	9	0	−13
India	1	2	3	0	7	4	0	×	7	0	−24
Asia	3	1	2	0	1	2	13	2	×	13	−38
Australia	0	0	0	0	0	0	1	0	4	×	−6
Total	122	41	245	2	20	28	134	81	239	20	933

Production can be expanded provided it makes economic sense for oil operators to do so. The actual investment is highly dependent on the type of oil produced. Light conventional oil usually requires lower investments than deep-water, arctic, or most unconventional oils. In addition, many exporting countries use part of the economic rent generated to subsidize other areas of their economy. This is the case in Middle East and in Russia, where a price of the barrel as high as one hundred dollars is required to balance the respective government's budget. Taking this into consideration, the investments that these countries can afford become a matter of careful selection of priorities.

Figure 3.16 maps conventional and unconventional total retrievable reserves volumes per region (© OECD/IEA, WEO 2012), as well as the corresponding incentive cost, the price at which it becomes relevant for an operator to start developing new fields (Advisor Perspectives 2015). The cost of unconventional oil development is considered in many regions where data is missing as equivalent to the one in the United States (this assumption remains to be proven). Also mapped on the graph are the government's budget breakeven price levels for the most important exporting countries (Knoema 2014; Energy Matters 2014).

An analysis of the graph shows that, at a price of 50 USD/bl (September 2015 baseline), Middle East countries are essentially incentivized to develop new fields. Some fields in Russia and Africa could as well be developed at this oil price level. There is a disconnect between sources on the incentive costs for shale oil in the United States, with price ranging between 60 USD/bl (BNP 2015) and 80 USD/bl (Advisor Perspectives 2015). The shale oil industry in the United States is nascent and has been growing extensively, with limited attention to efficiency thus far. Given the very short payback period and the fact that shale oil industry in the United States is made up of many small players, it is extremely likely that the level of

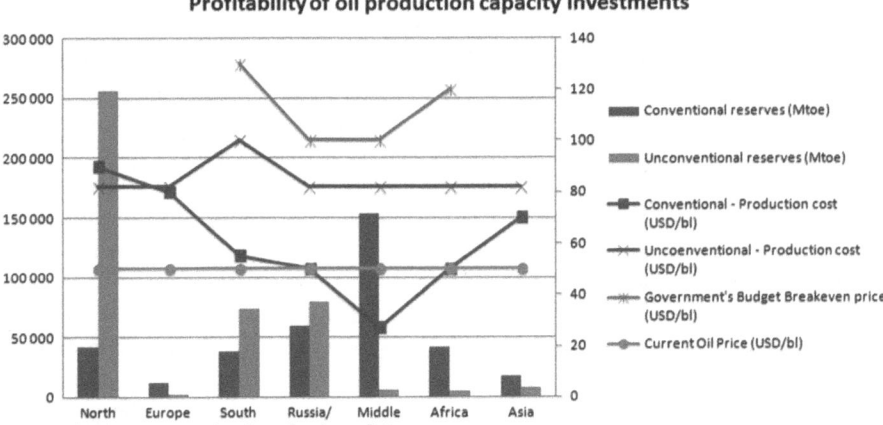

Fig. 3.16 Profitability of oil production investments (Advisor Perspectives 2015; Energy Matters 2014; © OECD/IEA, WEO 2012; Knoema 2014)

incentive costs will go drastically down in the coming few years, thanks to the triple effect of profitable wells' selection, industry consolidation, and the development of efficiency measures.

If many conventional fields can still be developed with a relatively low oil price, the governments' budget breakeven price (Russia, Middle East) is much higher than the actual oil incentive production cost. This puts the balance of the budget of these countries at risk, and may hamper the further development of new fields in those regions to a certain extent because of arbitration between long-term development of production capacity and the distribution of social benefits.

In the end, the current oil price level, if a result of the current imbalance between offer and demand, shall rise up again progressively, in order to fund the necessary investments that will be required to sustain the likely growth of oil consumption in the coming years. It is though very unlikely that oil prices get back to the historical levels reached in the past 10 years. All the experts tend to believe they will recover towards 60 USD/bl in 2016 and probably higher later on (Knoema forecast 2015; Advisor Perspectives 2015; Makortoff 2015; Sharma 2015; BNP 2015). This should be enough to trigger the necessary investments to meet the rising demand.

3.1.7 Summary

The oil market represents 43% of world trade, and over one third of global energy production. Resources and production remain concentrated in a few large regions: North America, Middle East and Eurasia (mainly Russia and Kazakhstan). Within the next 20 years, the oil market will continue to concentrate around these regions. Their share has already gained five points in the last 10 years and the share gain should continue, notably for the Middle East (+4pt) and North America (+2pt). While production concentrates, consumption will diversify, essentially in new economies which share has risen by more than nine points in 10 years. Moreover, oil consumption is shifting from North America and Europe to Asia. The geopolitics of oil is based on the relative independence of North America on one hand, and its strong ties with the Middle East, Europe and Asia. Finally, the situation of Russia remains open. Russia has long been a major energy partner of Europe. The growth of energy consumption in Europe is negative, and the continent is trying to diversify away from Russia. As a result, Russia faces difficulties in supporting the considerable investments that it needs to make to sustain production capacity in the long term. On the other side of the equation, China, with its tremendous needs for oil, offers an opportunity for Russia to develop its oil reserves.

The oil market has both short-term and mid-term cycles. In the short term, oil price is a balance between offer and demand. In the mid term, oil price is more of an indicator which triggers (or not) investments in additional production capacity. Today, short-term prices are more volatile than they have been in the past. The reasons for this are an increase in unplanned production disruptions, and the reduced capacity of OPEC countries to control their production output (notably upwards). In the mid term, oil prices should recover from the significant drop experienced in 2014/2015, in order to allow diversified investments, notably in the Middle East and in the United States. The shale oil industry in the United States is maturing. As a consequence, the incentive cost will likely be reduced in the coming years, making shale oil from the United States a strong contender on the oil market. This is clearly the new paradigm that the oil industry has entered. A major player has emerged. It will contribute heavily to the mid-term balance between offer and demand and to fixing price levels. The traditional paradigm that oil prices are determined by the production quotas of OPEC countries is thus far from being obvious. On the contrary, the strong flexibility of shale oil production in the United States could serve in the mid term as the main instrument to balance offer and demand and eventually control the overall market's price level. All experts seem moreover to have acknowledged this fact as they estimate the mid-term oil price to range around 70–80 USD/bl, which corresponds to the breakeven price of shale oil in the United States. The ability to control prices seems thus to have moved out of the Middle East. The consequences of this change remain to be written, especially when one considers that the traditional producers (Middle East, Eurasia) need prices as high as 100 USD/bl to balance their national budgets.

3.2 Coal: The Energy of New Economies

Of the energy resources available on Earth, oil is the primary one. It is used to propel automobiles and airplanes and used to manufacture things that have improved our day-to-day life. Coal is different. It does not benefit from a modern image. It invokes the sad images of the Industrial Revolution of the nineteenth century. Also, it is easy to produce and its reserves are relatively diversified. As a consequence, coal is never at the core of international discussions or geopolitical conflicts. From an energy point of view, though, it is as important as oil. From an environmental point of view, coal is a key resource . . . and a key issue.

3.2.1 The Chinese Market

Worldwide the amount of coal produced in 2013 was 3880 Mtoe. This corresponds to 95% of oil production (BP 2014) (Fig. 3.17).

Almost half of the world's coal is produced in China, which has a considerable appetite for it. India, Australia, Japan, South Korea, Taiwan and Indonesia are also large producers.

The evolution of coal production in the last 10 years shows how China dominates the market. Its growth represented 71% of the total increase in coal production. The increase in China was 7% per year, versus 4.2% worldwide (Fig. 3.18).

Its huge production capacity gives China the means to remain independent as it produces 96% of what it needs. This percentage was 105% in 2003, which means that China's energy independence has slightly eroded over time. Indeed, the fast

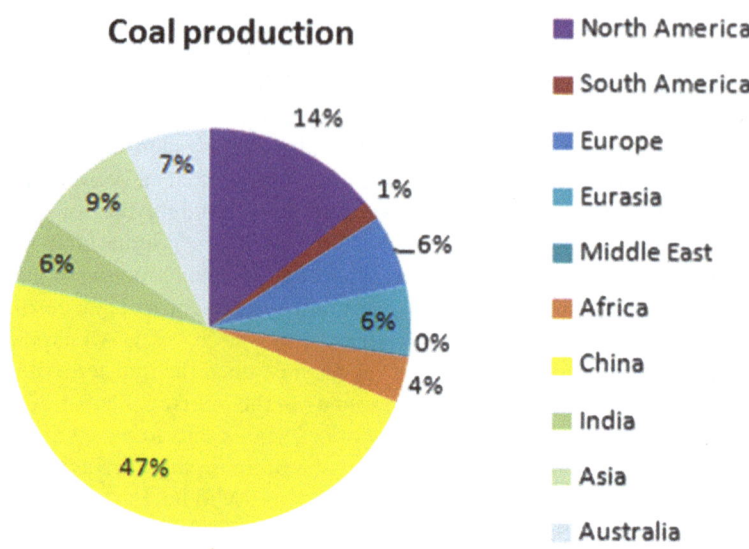

Fig. 3.17 Worldwide coal production (BP 2014)

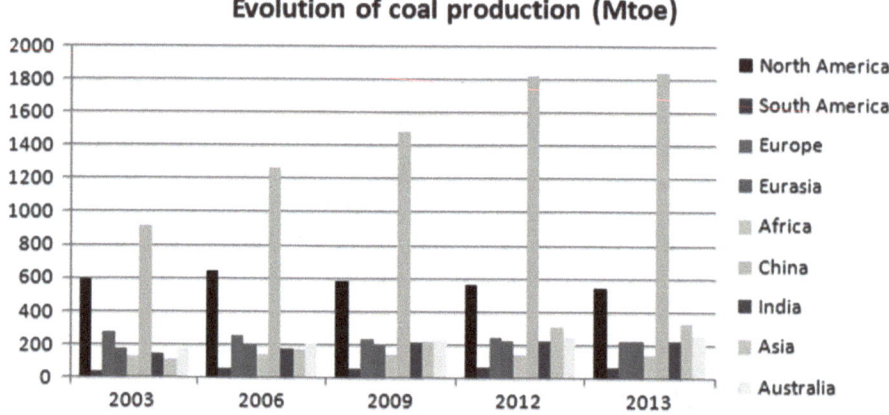

Fig. 3.18 Worldwide coal production per region (BP 2014)

growth of its economy led China to more than double its consumption of coal in the last decade.

In Asia (excluding China), coal production has increased by 11% on average, much more than its consumption (3.7% on average). While the region imported 60% of its coal 10 years ago, this has fallen to 20% today. India is also a net importer of coal; it imports around 30% of its coal needs. Its own production does not meet the growth in consumption, which tops 7% per year on average.

Fig. 3.19 Coal usages
(© OECD/IEA, WEO 2012)

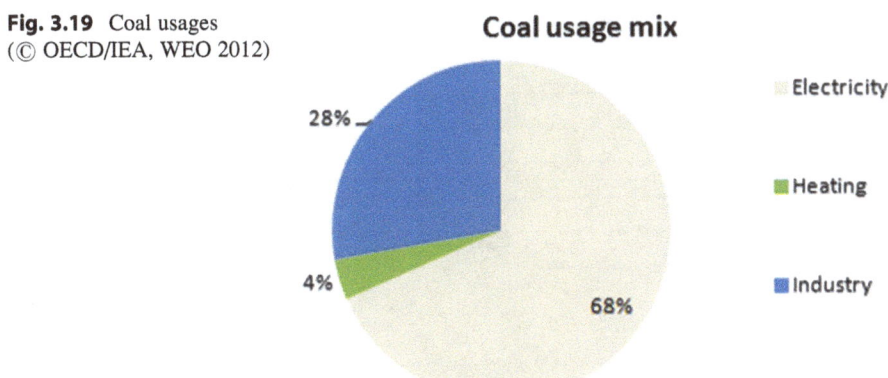

Europe imports 40% of the coal it consumes. This percentage slightly decreased, essentially from reduced consumption, which has steadily decreased by 1% per year on average. North America is both a key producer and consumer of coal. In the last 10 years, it became a net exporter of coal. Indeed, production slightly decreased, but consumption reduced even faster, around 2% per year. North America produces today around 10% more coal than it consumes.

Coal is mainly used to produce electricity (68% of the total usage). The growth in the demand for electricity thus has a direct impact on coal consumption (Fig. 3.19).

In summary, coal is the energy source that posted the highest growth in the last 10 years (4.2%), far above oil (1%) or natural gas (2.5%), despite its reputation as being in decline. Coal consumption only represented 70% of oil's consumption 70 years ago, compared to 95% today. The growth in demand for this resource is essentially related to the economic transition in Asia, in particular China. As coal is essential for electricity production, it supports the economic transition of these regions. It is thus at the heart of the world energy transition.

3.2.2   Electricity Production Drives the Growth of Coal

Electricity production should grow by more than 70% in the world by 2035, and will correspond to 70% of the total growth in coal consumption, according to the International Energy Agency (2012). This will correspond to a yearly increase of 0.8% per year on average for coal. Forecasts differ however from one source to another. Shell (2016) estimates the growth of coal to be positive, and to vary between 1.5% per year and 2.1% per year depending on the scenario. Exxon Mobil (2016) estimates coal production evolution will be flat, while Statoil (2016) estimates a negative growth of coal in two scenarios out of three. Main uncertainties relate to actual public measures to force the substitution of coal power to renewable, as well as the ability of renewable power to reach true competitiveness. The scenario of the International Energy Agency presented here is in between those different scenarios (Fig. 3.20).

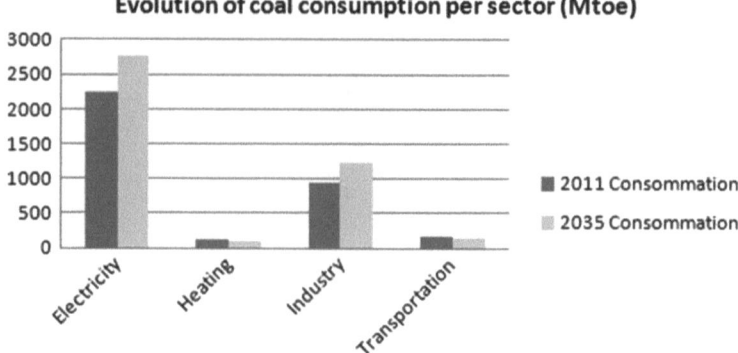

Fig. 3.20 Evolution of coal consumption per sector (© OECD/IEA, WEO 2012)

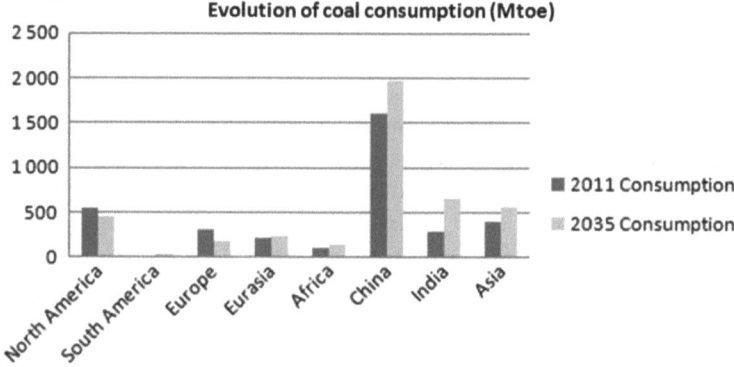

Fig. 3.21 Evolution of coal consumption per region (© OECD/IEA, WEO 2012)

This growth will essentially come from Asia, as demand for coal in North America and in Europe is in decline. Demand in China should slow down too, after growth levels above 7% per year for the last 10 years. According to the International Energy Agency (2012), coal production in the country is not expected to exceed 1% per year. At the same time, the growth of electricity production is estimated to be around 5% per year on average. This means that China would soon modify its energy policy and its electricity production mix.

India should increase its production and consumption of coal by around 3.6% per year on average, and the rest of Asia around 1.5% (Fig. 3.21).

3.2.3 Coal Reserves

In 2013 coal reserves corresponded to an equivalent 623 billion tons of oil equivalent, or 160 years of consumption at the actual pace. This is three times more than conventional "proven" reserves of oil. These reserves are well distributed across the world, except maybe for South America, which possesses only 2% of the world's reserves. North America holds the top position with 28% of the world's reserves, with Russia second with 21%. Europe has 13% of the world's reserves, with Germany accounting for more than half. About 13% of the world's reserves are in China (Fig. 3.22).

North America reserves correspond to 300 years of production at the current pace. Australia, a large producer, can continue producing at this pace for another 200 years. Russia and Eurasia as a whole have considerable resources which they do not capitalize on as they mainly use natural gas.

Finally, China is by far the top coal producer and consumer in the world. Only a couple of decades of coal production remain domestically, however, after which China faces a major energy mix disruption. When coal reserves are depleted, the country will have to rely on other sources of energy or will have to import coal in the vast quantities it consumes (Fig. 3.23).

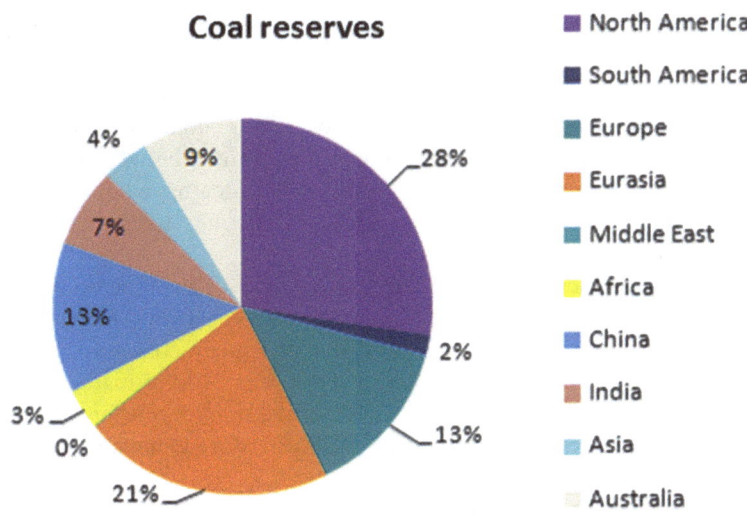

Fig. 3.22 Coal reserves (BP 2014)

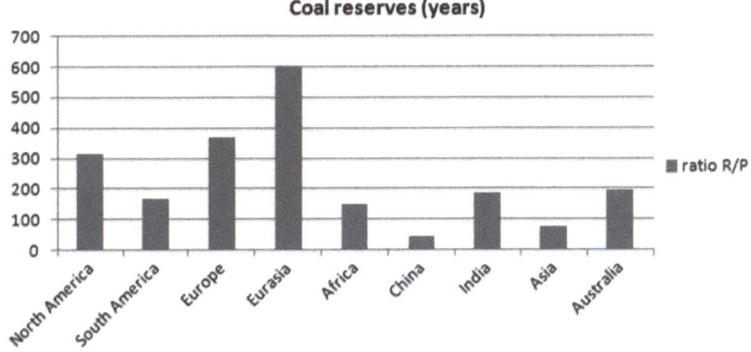

Fig. 3.23 Coal reserves in years of production (BP 2014)

3.2.4   The Coal Market

The price of coal follows Ricardo's theory more accurately than the price of oil, which we have seen is dependent on a multitude of other factors. The marginal cost weighs more heavily in setting the overall price of coal on the global commodity market. This is because the amount of taxes and subsidies are less important for coal than for oil. As well, the fact that coal resources are more scattered across the globe have lessened the volumes of coal traded globally and simplified to a certain extent the chain of cost and benefits' allocation among the various stakeholders, leading to an overall price which reflects better the actual marginal cost of production. The cost of transportation is also important with regards to the cost of production. Therefore two distinct markets for coal coexist: the Atlantic market and the Pacific market. These two markets have distinct prices and different dynamics, and there are not many exchanges between them—trade between the two markets does not exceed 15% of the total volume supplied (© OECD/IEA, Coal Report 2012) (Table 3.3).

The price structure and the operating model also differ between the Atlantic and Pacific markets. In the Atlantic market, the "spot" market dominates, whereas in the Pacific market, the coal market features long-term contracts between producers and consumers. Large steel companies from Japan and Korea have so far managed to prevent the emergence of a "spot" market, privileging long-term and stable relationships with their suppliers.

The coal market is less developed and less speculative than the oil market. It is dominated by four main companies—Rio Tinto, Glencore, Amcoal and BHP Billiton—that, together, own more than 50% of the steam coal market. The coke market is dominated by BHP Billiton with more than 30% of market share. It is followed by Fording, which has less than 6% of the market (Furfari 2007). In the Atlantic market, Glencore, Amcoal and BHP Billiton dominate. Rio Tinto is essentially present in the Pacific market.

The coal market is finally less developed and less speculative than the oil market. Few large companies ensure the majority of the international trade. Now, the coal market remains in the end dominated by China which represents around 50% of the world market. This is as a result a much more fragmented market than it seems.

Table 3.3   Worldwide coal exchanges (© OECD/IEA, Coal Report 2012)

Coal (Mtoe)	Transit energy to		
Transit energy from	Atlantic region	Pacific region	Total
Atlantic region	158	58	216
Russia	59	34	93
South Africa	29	42	71
Pacific region	45	602	647
Total	291	736	1027

3.2.5 Summary

The coal market doubled in size in 10 years. However, most of this growth is related to China, which today represents almost half of global production, far ahead of the rest of the world. The growth in production in China has represented 71% of the total world coal production in the last 10 years. China supported its economic development by doubling its coal production capacity. Now, China produces 96% of its own consumption, ten points less than a decade ago. This situation should continue to worsen in the coming 20 years and China will not be able to continue to support the development of its industry and of its electricity production with coal only. The country thus needs to look for alternative energy resources. The rest of Asia is also extremely dependent upon coal. Although the coal market seems smaller than the oil market, coal has a significant role in the energy mix worldwide. Indeed, coal production corresponds to 95% of oil production. The market is simply more local and more fragmented than the oil market. Coal is being used to produce electricity, and therefore should strongly benefit in the decades to come from the large increase of electricity demand, notably in Asia. The increase in electricity production should represent 70% of the growth of coal production in the next 20 years. Coal production can thus be expected to continue to grow, as it supports the economic transition of the new economies. It should however grow at a much slower pace in the coming years than what it used to, due to growing alternatives such as natural gas and renewable energies.

3.3 Natural Gas: Star Product of the Twenty-First Century?

Primary energy production is largely based on oil and coal. Next comes natural gas, which represents almost a quarter of primary energy production worldwide. The average growth of natural gas production in the last decade has been 2.5 times higher than oil, around 2.5% per year on average, but slower than coal (4.2% per year). However, if we exclude China, the growth of coal production has averaged 2% per year. Natural gas is therefore one of the most if not the most dynamic source of energy today.

3.3.1 The Various Usages of Natural Gas

The strength of natural gas lies in the variety of its applications. It can be an alternative to both oil and coal in electricity production. It can also replace oil or coal for heating systems, its main use in OECD countries. Finally, it can substitute oil or coal in a variety of industrial applications: as fuel for a heating process or as feedstock for specific applications. It can even serve as an oil substitute for transportation, even though such usage is today very limited (Fig. 3.24).

Natural gas is therefore an alternative energy by nature. If it does not dominate any application, it is used in most sectors with every time a range of specific advantages. This is the diversity of these usages which explain the dynamism of its development.

North America is the biggest consumer of natural gas in the world, with 28% of global consumption. The particularity of North America is that it ensures its own production of natural gas. There are indeed almost no exchanges between North America and other regions of the world. This can be simply explained by the fact that the cost of transporting natural gas is extremely expensive. Asia comes second with around 20% of consumption worldwide, and Europe third with 17%. Eurasia with around 190 million inhabitants comes fourth with 15%, which makes the region the biggest natural gas consumer per individual; natural gas consumption in Russia is 3 toe/year/individual a year. In comparison, North America, Europe and the Middle East come in at around 1–1.5 toe/year/individual (Fig. 3.25).

The consumption of natural gas has strongly evolved over the last 10 years. It decreased in Europe by 0.5% per year on average, and increased in the Middle East, Africa, South America and Asia at a growth rate of about 5% per year (versus 3.5% per year for oil). In North America, it progressed by 1.7% per year on average. There, natural gas was the only source of energy not to decline in terms of consumption; consumption declined by 0.8% per year on average for oil and more than 2% per year for coal. In North America, natural gas gained market share from other energy sources.

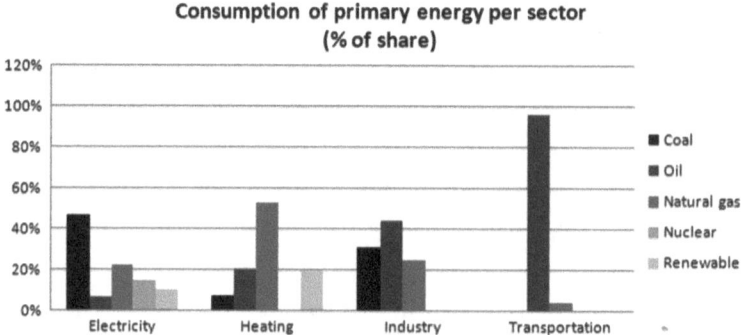

Fig. 3.24 Consumption of energy per sector (© OECD/IEA, WEO 2012)

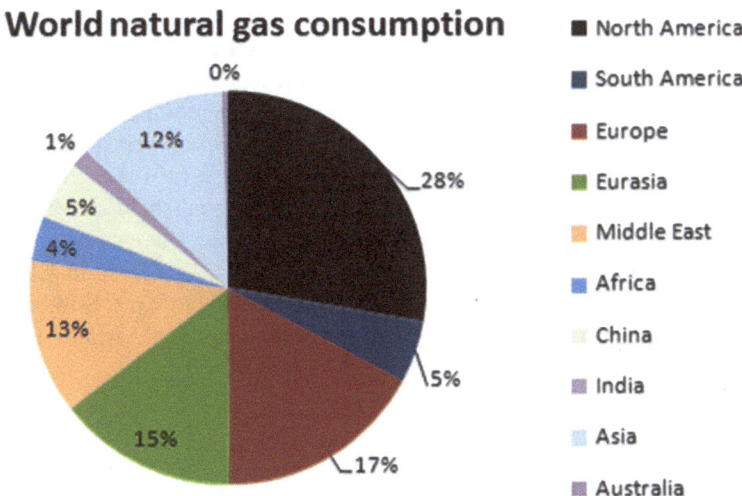

Fig. 3.25 Worldwide natural gas consumption (BP 2014)

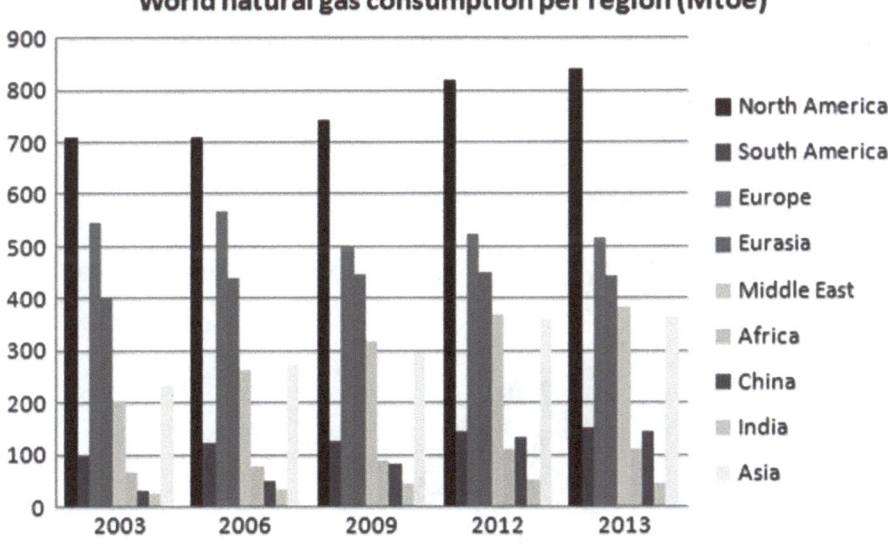

Fig. 3.26 Evolution of natural gas consumption per region (BP 2014)

The most striking element of the global natural gas market is the growth in consumption of the resource in China. Although the country is a comparatively small consumer, consumption there has increased by an average of 17% per year in the last few decades. At the same time, consumption of oil grew by 6% and that of coal by 8% on average. The share of natural gas thus increased within the Chinese energy mix. Despite this growth, the increase in gas consumption in China still represents less than 10% of the total increase in energy consumption in the country. Natural gas is a credible alternative to coal for electricity production and a very interesting option for space and home heating, provided the necessary distribution infrastructure is in place. It is easy to guess why China is moving this path. Russia's huge reserves of natural gas are indeed geographically very close to China. Furthermore, according to the Energy Information Agency (2014) estimates, China would have the world's highest reserves in unconventional shale gas. The International Energy Agency (2012) estimates that, by 2035, the share of unconventional gas could represent 75% of gas consumption (about 225 Mtoe) in China. This would correspond to a production output twice as large as today's. Still, natural gas production would remain much smaller than that for coal, which today tops 1800 Mtoe (Fig. 3.26).

3.3.2   Natural Gas Production Concentration

North America, Russia and the Middle East are today the main natural gas producers in the world (Fig. 3.27).

The share of Middle East gas jumped from 10 to 17% of the world energy production over the last 10 years. North America's remained around 27% in percentage while volume increased. Eurasian production maintained volume even as it declined in percentage terms from 24 to 22% of global gas production. In Europe, production dropped from 14 to 9% (Figs. 3.28 and 3.29).

A concentration of gas production in the Middle East and North America can thus be observed. The output growth in Middle East has been somewhat detrimental

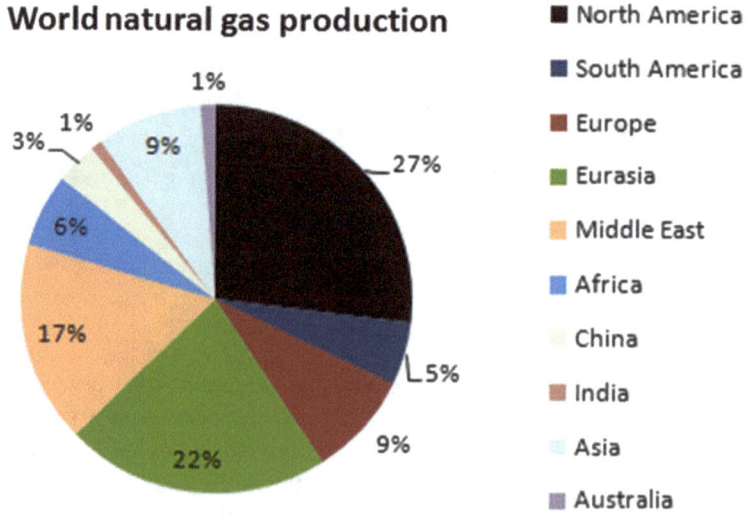

Fig. 3.27 Worldwide natural gas production (BP 2014)

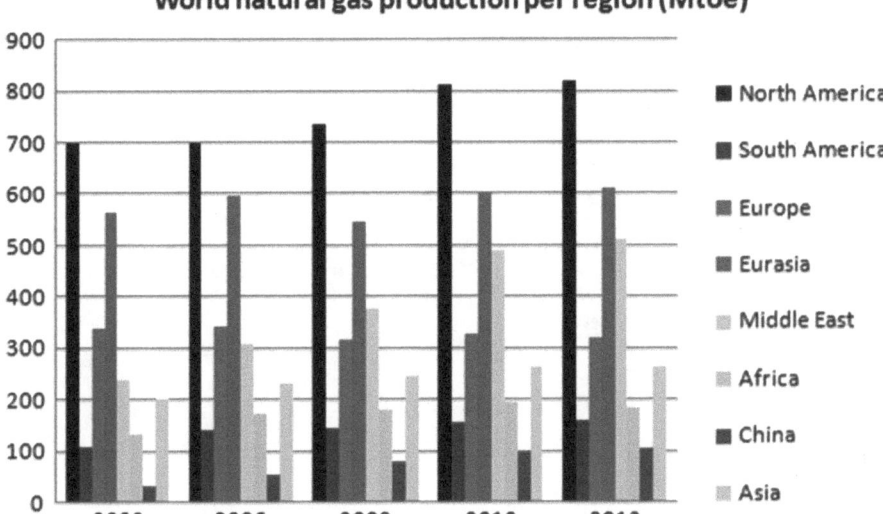

Fig. 3.28 Evolution of natural gas production per region (BP 2014)

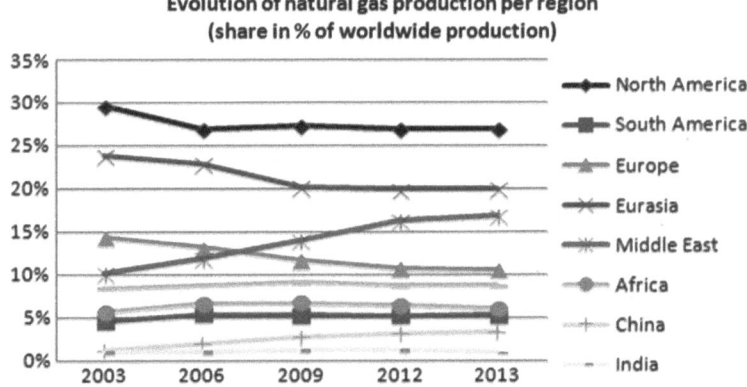

Fig. 3.29 Share of natural gas production per region (BP 2014)

to Russia, which share has dropped in the last decade. In North America, growth in volume was mainly fueled by increased local consumption, with 1.6 toe/year/individual consumed on average, one of the highest in the world (after Eurasia).

3.3.3   Geopolitics of Natural Gas

The volume of exchanges of natural gas across regions is much smaller than for oil (BP 2014), corresponding to just 18% of exchanged oil volumes (Table 3.4).

Table 3.4 Worldwide natural gas exchanges (BP 2014)

Gas (Mtoe) Transit energy from	Transit energy to										
	North America	South America	Europe	Russia/Eurasia	Middle East	Africa	China	India	Asia	Australia	Total exports
North America	×	0	0	0	0	0	0	0	0	0	0
South America	5	×	3	0	0	0	0	0	3	0	11
Europe	1	3	×	2	0	0	0	0	1	0	7
Russia/Eurasia	0	0	149	×	5	0	25	0	13	0	191
Middle East	3	2	29	1	×	0	10	14	68	0	126
Africa	1	2	46	0	1	×	2	1	17	0	69
China	0	0	0	0	0	×	×	0	0	0	0
India	0	0	0	0	0	0	0	×	0	0	0
Asia	0	0	0	0	0	0	9	0	×	6	15
Australia	0	0	0	0	0	0	4	0	23	×	27
Total imports	10	6	228	2	6	0	49	16	124	6	447
Consumption	839	152	515	443	386	111	146	47	366	16	3021

The main drawback to natural gas exchange is the prohibitive cost of transportation and distribution. Together, these costs can add up to 60% of total production cost. Natural gas is traditionally transported through gas pipelines, which are extremely expensive and are very complex to operate and maintain, especially in sensitive areas (e.g., Middle East, Eastern Europe). Liquefied Natural Gas (LNG), which is transported by sea, now corresponds to a third of total exchanges across regions (BP 2014). The LNG market is accelerated by rising demand from Asian countries, which leads the Middle East, Africa and Australia in building large infrastructures to ship out LNG.

Overall, international exchanges of natural gas remain limited and represent only 15% of consumption worldwide. One third of these exchanges correspond to distribution of Russian natural gas to Europe, while another third corresponds to Middle East exports to Asian countries. This is in the Middle East that the LNG market develops the fastest.

The vast majority of natural gas remains however produced and consumed locally, within the region itself. Natural gas is therefore essentially a regional market.

3.3.4 Natural Gas: A Serious Alternative to Coal

The natural gas market should grow by an average of 1.7% a year for the next 20 years (© OECD/IEA, WEO 2012), three times faster than oil but less than the 2.5% growth in the last 10 years. As with oil, several scenarios were developed by the International Energy Agency; the intermediate one is selected for discussion here. Exxon Mobil (2016) presents a similar scenario with around 1.7% growth for the coming decades. Shell (2016) considers the growth of natural gas to be higher, between 1.9% ("Oceans" scenario) and 2.5% growth per year ("Mountains" scenario). Statoil (2016) considers the growth to be lower, between 0 and 1.1% per year depending on the scenarios. As natural gas is a credible alternative to coal for electricity production, production is expected to grow everywhere, with the exception of Europe. North America shall continue to remain independent. Russia is planned to grow at 1.5%, twice the growth of its consumption. The country is likely to compensate for the drop of exports to Europe by exporting to China. To facilitate such exports, investments need to be made in both production capacity and transportation infrastructure. Bilateral agreements with China could help ramp up operations in Siberia and in the Far East, to the benefit of China. Another large producer is the Middle East, where production is expected to grow by more than 2% per year on average. Output growth in Africa is estimated at 3% per year. These regions will supply both Europe and Asia. The growth of production in Asia (5% per year in China, 3% per year in India, 2% per year in the rest of Asia) would not be enough to compensate for the growth in consumption in these countries. Consumption is expected to expand at more than 7% per year in China and more than 5% per year in India (Figs. 3.30 and 3.31).

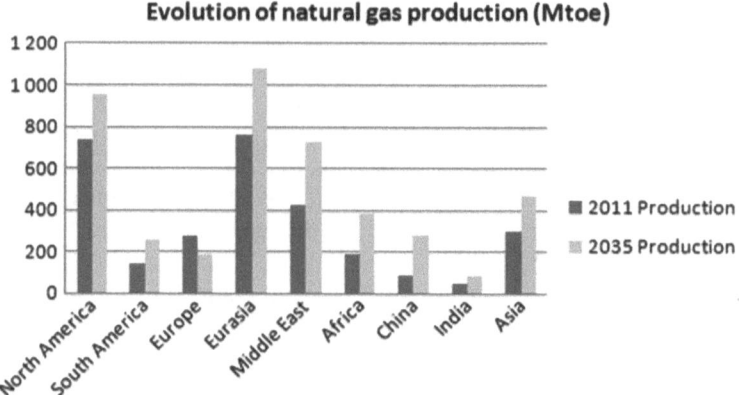

Fig. 3.30 Evolution of natural gas production (© OECD/IEA, WEO 2012)

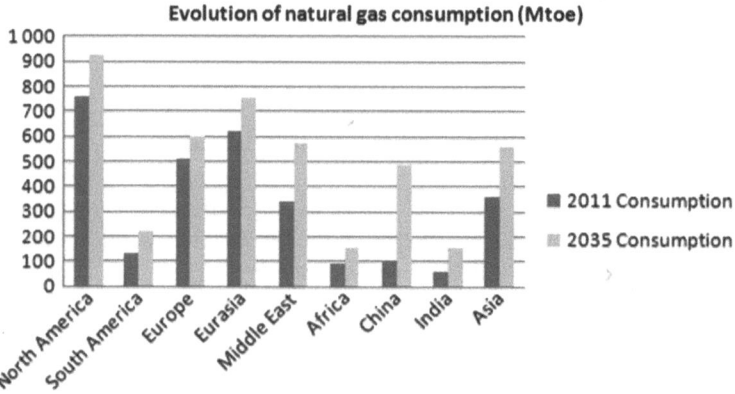

Fig. 3.31 Evolution of natural gas consumption (© OECD/IEA, WEO 2012)

The expansion of the natural gas industry will be driven by electricity production, with growth for that application estimated at above 1.8% per year on average over the next 20 years. District heating should not grow by more than 0.9% per year on average. Most of this growth will come from China, which will invest in the sector. The market in Europe and North America is fairly stable and saturated. Industrial applications should also grow fast, around 2.9% per year on average, even though the base is low (16% of total). In those applications, natural gas will progressively be used as an alternative to coal for heating processes as well as a raw material for some energy processes (e.g., in refineries) (Fig. 3.32).

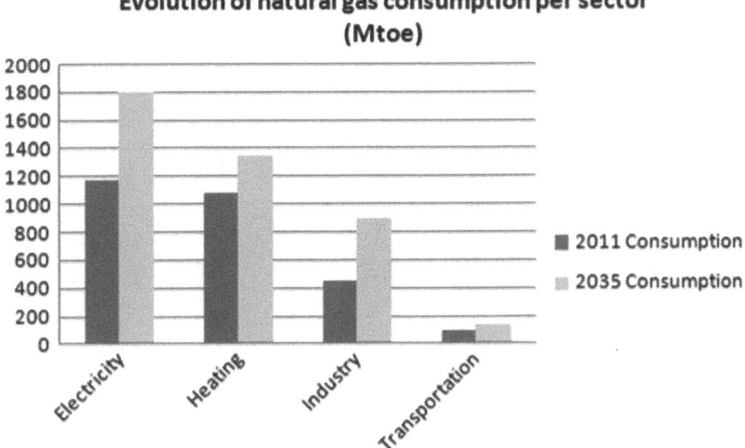

Fig. 3.32   Evolution of natural gas consumption per sector (© OECD/IEA, WEO 2012)

3.3.5   Natural Gas Reserves

"Retrievable" natural gas reserves worldwide correspond to 800 TM^3, or a bit more than 700 billion tons of oil equivalent. This represents 200 years of consumption at the current rate [3.3 TM^3 per year in 2012 (BP 2014)]. "Proven" reserves equal 190 TM^3 (BP 2014) or around 168 billion tons of oil equivalent, which is 60 years of production at the current pace. The reserves increase year after year, with the Middle East and Russia being the main regions where "proven" reserves can be found. However, when looking at the total footprint of "retrievable" natural gas reserves, reserves are spread across the globe in a more balanced manner, notably in Asia Pacific and, in particular, in China, which has the highest world reserves of unconventional gas (Carto 2014; EIA 2013). A number of regions also have important "retrievable" reserves. This supports the assumption that natural gas could become one of the main energy sources of the twenty-first century. These reserves however are not yet all being developed, which explains the gap between "proven" and "retrievable" resources (Figs. 3.33 and 3.34).

In North America, the spectacular rise of shale gas production already confirms the continent's energy autonomy. China faces gigantic needs in terms of energy resources and is already drawing down its coal resources. The discovery of unconventional gas reserves in China represents a historical opportunity for the country, provided that it is economically sound to develop the reserves.

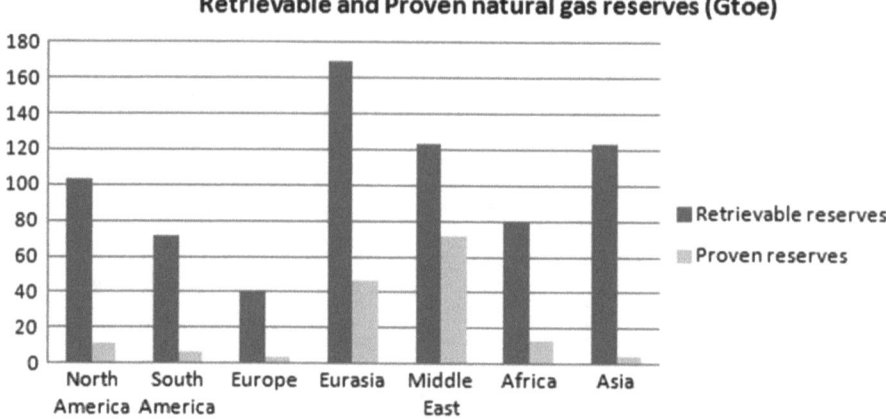

Fig. 3.33 Natural gas reserves (Carto 2014; EIA 2013; BP 2014)

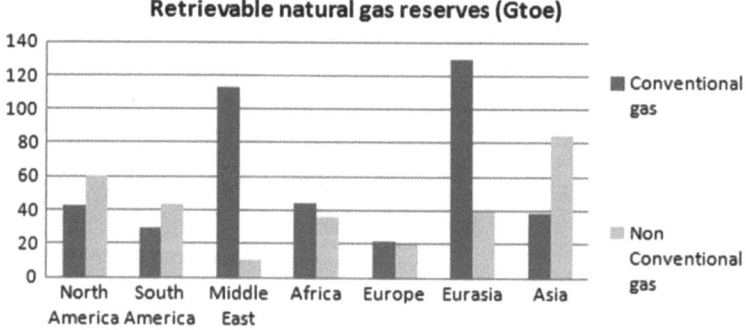

Fig. 3.34 Retrievable natural gas reserves (Carto 2014; EIA 2013; BP 2014)

3.3.6   Summary

The natural gas market is above all a regional one, with less than 15% of production shared across regions. It is as well a market that requires significant investment in transportation infrastructure, whether these be gas pipelines or LNG ports. The LNG market represents only 30% of inter-regional exchanges, less than 5% of global production. The vast majority of natural gas therefore transits through gas pipelines, either within a region, or across regions. Such infrastructure is very expensive to build, operate and maintain. This drives the structure of the gas market. The market is indeed built around long-term contracts which guarantee to the purchaser the receipt of product and to the supplier the profitability of its investment over time.

Clearly natural gas is an alternative source of energy, whatever the sector in which it is used. Its price is therefore influenced by the fuels that it is able to replace: oil and coal. In practice, natural gas prices vary between those for oil and coal.

Increasing natural gas production in the Middle East and Africa in the next 20 years (and, to a lesser extent, Asia and Australia) should boost market fluidity by increasing LNG volumes. No major market disruptions are foreseen in the market, and volumes transiting through pipelines should remain dominant. The structure of long-term contracts should thus not be changed. Europe, which wishes to better diversify its procurement in order to lessen its dependence on Russian natural gas, could open its market to the Middle East, Africa, or even North America; this would create the conditions for a more open market for natural gas.

Russia and China are expected to partner up to increase shipments of gas. Beyond these market factors and changes, the emergence of unconventional gas will to a certain extent change the geopolitical balance across countries. Shale gas has already confirmed the total energy autonomy of North America. China holds the most world reserves in unconventional gas and this will prompt the country to develop the capabilities to produce natural gas for use as an alternative to coal. While this will help China to meet its growing energy needs, it would probably not be enough to compensate for coal consumption, which reserves are progressively being depleted. The share that shale gas would take in the mix will also depend on how economically relevant it will be to develop these resources compared to alternative options.

3.4 The Spectacular Growth of Electricity Production

Electricity production is at the heart of the energy transition. While it represents 38% of primary resources consumption, it accounts for only 20% of final energy consumption. This is due to the high level of energy losses incurred when transforming primary resources into electricity.

In electricity production, fossil fuels (oil, coal, natural gas) are combusted. The combustion heats up a coolant that then expands. Such expansion results in an increase in pressure that drives mechanically the rotation of a turbine. The mechanical rotation is transformed into electricity thanks to an alternator. From start to finish, calorific energy is transformed into electrical energy through a mechanical conversion. The multitude of the transformation at each of the various steps and the associated losses make electricity production an inefficient process. On average, 60% of the calorific energy is lost in the transformation.

Nuclear energy is produced in like manner. Instead of combustion as for fossil fuels, it is the nuclear fission reaction that creates massive emissions of heat. Most renewable energies operate under different principles. Hydro-electricity uses the potential energy of moving water. Once the water is released from a dam, it gains speed and is used to directly rotate turbines within the dam. Wind farms use the kinetic energy of the wind to rotate turbines. Photovoltaic energy uses solar radiation which produces electricity by actuating semiconducting materials. Finally, geothermal energy or concentrated solar plants operate in a more traditional way as they use heat to expand a fluid and create mechanical pressure. Geothermal energy harnesses the natural heat deep inside Earth, while concentrated solar uses the heat of the sun.

One of the main characteristics of electricity is that it cannot be stored. This means that production needs to balance out consumption at all times. Excluding losses in distribution, production always equals consumption.

3.4.1 Overview of Electricity Production Worldwide

Electricity production in 2010 amounted to 18,000 TWh, or about 1500 Mtoe. This corresponds to what was finally consumed. The primary resources required to produce this amount of energy corresponded to 4800 Mtoe (© OECD/IEA, WEO 2012), 38% of total primary resources consumption.

North America and Europe accounted for 43% of the global demand for electricity; China had a 19% share (© OECD/IEA, WEO 2012). On a per capita basis, North America consumed the most electricity, followed by Russia and Europe. China came in at around the world average of about 0.2 toe/year/individual, which is four times less than in North America and twice less than in Europe (Fig. 3.35).

Increasing urbanization and economic vitality in new economies lead to significantly higher electricity consumption there. There is thus an impending upheaval in Asia (and then Africa) with regards to electricity consumption. When one looks at the huge consumption (and waste) of primary fossil resources, the "electrical" catch-up of new economies, because of its combined size, is a tremendous threat to the world's energy balance.

Electricity is primarily produced from fossil fuels (75%), mainly coal (46%). Oil finds less favor in electricity production (6%), though it is used for such in Japan and the Middle East. Nuclear energy represents only 15% of total electricity production, and is essentially used in OECD countries (Europe, North America, Japan) and in China. Renewable energy represents 10% of the electricity production worldwide, thanks notably to hydroelectricity (Fig. 3.36).

These percentages consider the amount of primary energy consumed overall in tons of oil equivalent to produce the necessary electricity output in TWh. Conventional power plants have yields between 30 and 40%. A great portion of the primary energy consumed is thus wasted. This is not the case for renewable energies. Theoretically, the transformation ratio of renewable energies is 100% since they are free and renew themselves. This means that the overall share of renewable

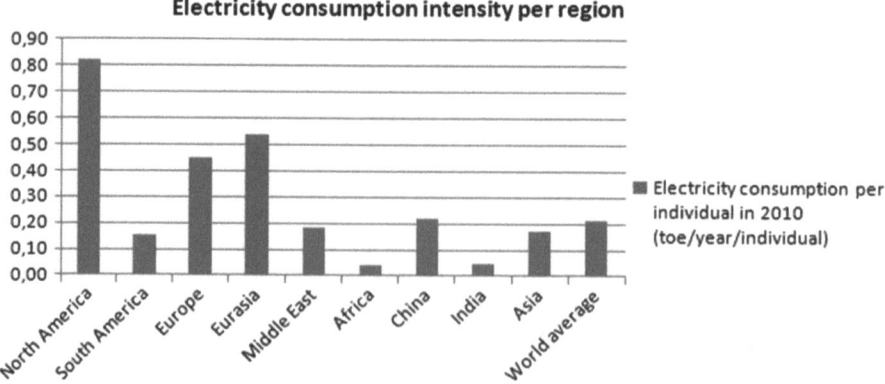

Fig. 3.35 Electricity intensity per region (© OECD/IEA, WEO 2012)

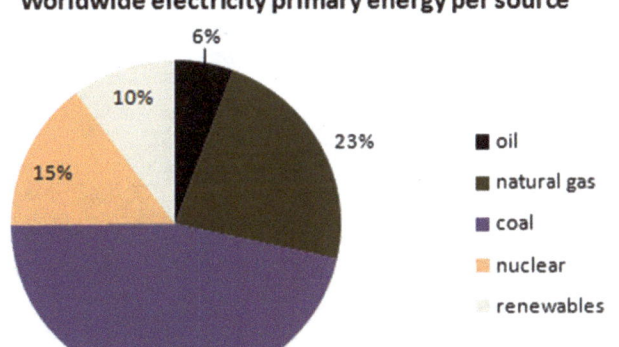

Fig. 3.36 Worldwide electricity primary energy per source (© OECD/IEA, WEO 2012)

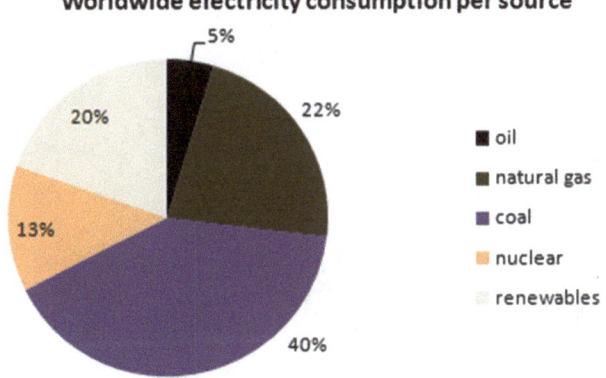

Fig. 3.37 Worldwide electricity consumption per source (© OECD/IEA, WEO 2012)

energies is actually higher from a demand standpoint. The share of renewable energies in the overall mix of final electricity consumption is closer to 20% overall (© OECD/IEA, WEO 2012). Hydroelectricity represented 81% of total renewable electricity generation in 2010 (© OECD/IEA, WEO 2012) (Fig. 3.37).

3.4.2 Medium-Term Perspectives

Electricity consumption is expected to grow by more than 70% in the next 20 years (© OECD/IEA, WEO 2012; UN/DESA 2014), the result of world population growth, rising living standards in new economies, and increased urbanization. Most available forecasts predict a similar evolution. ExxonMobil (2016) expects an increase of around 80%, Shell (2016) between 84 ("Mountains" scenario) and

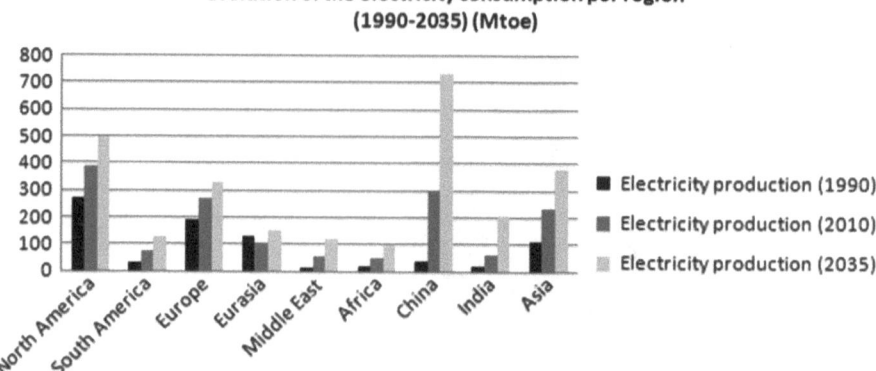

Fig. 3.38 Evolution of electricity consumption per region (© OECD/IEA, WEO 2012; UN/DESA 2014)

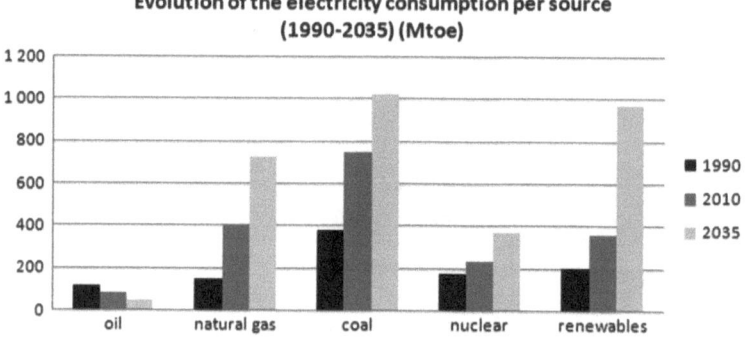

Fig. 3.39 Evolution of electricity consumption per source (© OECD/IEA, WEO 2012; UN/DESA 2014)

116% ("Oceans" scenario), and Greenpeace (2015) an increase between 70 ("Reference" scenario) and 81% ("Advanced Energy [R]evolution scenario). The growth of electricity consumption will be spectacular in the coming decades. The production of electricity is expected to increase in tandem with its consumption. Output should be on the uptrend in new economies, where demand is rising fast, mainly in Asia, particularly China. China is expected to account for 40% of production growth worldwide, and the country would in 2035 become the top electricity producer in the world with a 28% slice of global electricity production, ahead of the United States (Fig. 3.38).

Such growth across the globe will require all means available. While the combined share of fossil fuels is expected to decrease in percentage terms (from 67% today to 57% in 2035), they will keep growing in absolute value. Renewable energy will pick up a bigger share, and nuclear energy will remain stable at around 13% of total global electricity production (Fig. 3.39).

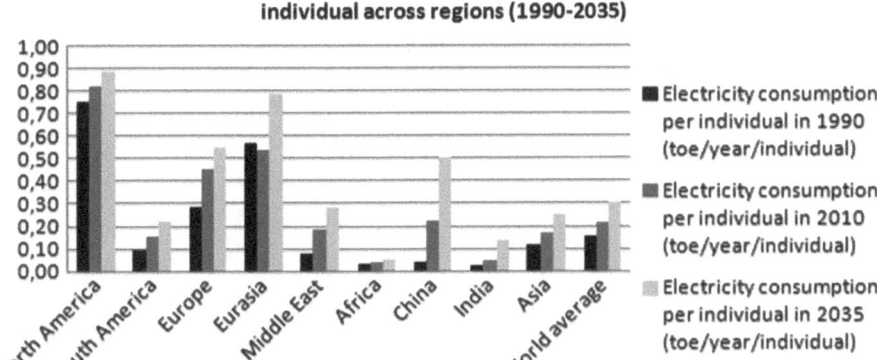

Fig. 3.40 Evolution of electricity consumption per individual (© OECD/IEA, WEO 2012; UN/DESA 2014)

The electricity consumption intensity will also vary across the world. China would reach Europe levels of around 0.5 toe/year/individual, which, considering China's population size, is the main factor of electricity production growth in the world. The United States should remain the most electricity-intensive country in the world with around 0.9 toe/year/individual, three times the world average. This evolution is expected to be more moderate in other regions of the world, in particular Africa, India and the rest of Asia (Fig. 3.40).

The spectacular growth of electricity production in the mid term can then be explained by two factors that combine together. On one hand, world population growth drives mechanically an increase in electricity production. The world population will increase by around 20% by 2035 and electricity production will increase by the same factor. On top of that, urbanization of new economies and the improvement of living standards, along with the development of the middle class in those economies, will lead to a massive increase in electricity consumption intensity. The electricity consumption per individual will rise from 0.21 toe/year/individual to 0.31 toe/year/individual in 2035. In absolute value terms, this means that the improvement of the living conditions accounts for 70% of electricity demand growth. This is particularly true in China, where just the improvement of living conditions should account for almost 40% of the total electricity growth in the world, more than the growth related to the increase in world population (Fig. 3.41)!

More than the growth of the world population, this is the improvement of the living standards in new economies, thru the development of an emerging middle class, which shall lead to a spectacular increase of the electricity demand. This should continue during the twenty-first century, as other economies achieve their economic transition and their complete integration in the global economy.

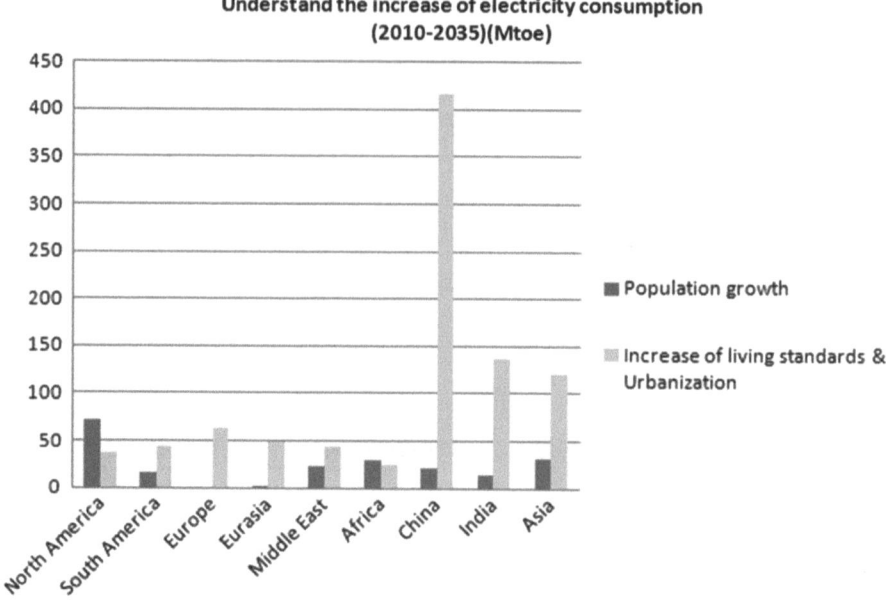

Fig. 3.41 Understand the increase of electricity consumption

3.4.3 Nuclear Energy: An Energy in Decline?

Nuclear electricity is an alternative to fossil fuels. While the yield of a typical nuclear power plant is not much better than that of a traditional thermal power plant, it uses a small amount of fuel to generate large quantities of heat. It also does not release any greenhouse gases to the atmosphere.

3.4.3.1 An Energy in Decline?

The International Atomic Energy Agency (2014) planned several possible scenarios for nuclear electricity production growth in the coming 20 years (Fig. 3.42).

The production of nuclear electricity in North America and Western Europe is likely to decrease, but production should rise at a fast pace in other regions. In China and India, the growth should be spectacular. Nuclear electricity production should rise by 6% in the low-base scenario to more than 9% per year on average in the high-base scenario, which would represent an additional 120 GW of installed capacity for those two countries.

Whether production grows (or not) in a country or region depends on policy decisions that influence the geography's energy mix. As an example, future production growth in Europe will depend on many factors:

- Some countries may choose to exit the nuclear market place

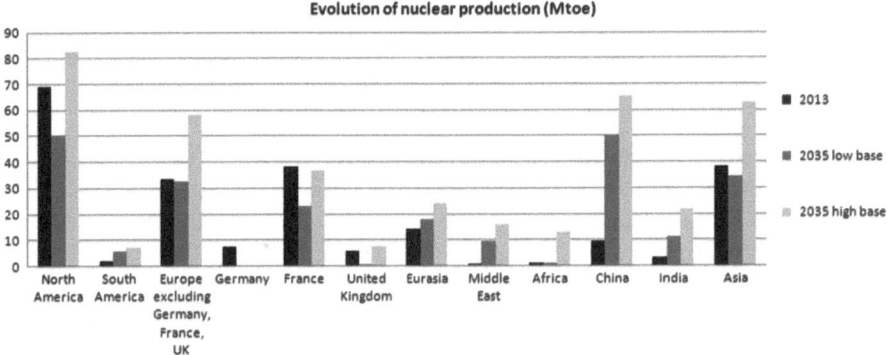

Fig. 3.42 Evolution of nuclear production (IAEA 2014)

- – France: −25 GW reduction if the government decides to exit progressively, −2 GW reduction if it decides to maintain capacity
- – England: −8 GW reduction or +4 GW increase
- – Spain: −5 or 0 GW
- – Sweden: −9 or −2 GW
- Some other countries may decide to develop further their nuclear capacities
  - – Italy might decide to actually develop nuclear energy (up to 13 GW by 2035).
  - – Countries in Eastern and Northern Europe will likely use more nuclear energy to reduce their dependence on natural gas from Russia: Ukraine (+9 GW or +14 GW), Poland (+7 or +10 GW from a zero base), Finland (+3 or +4 GW) and Czech Republic (+3 or +4 GW).

In Africa, the difference between the scenarios depends on the energy policy of South Africa, which could take its existing installed base from less than 2 GW to around 20 GW in 2035 in the high-base scenario.

A number of countries in Asia are considering developing their own nuclear power. Among them are Vietnam, Thailand, Malaysia and Indonesia. However, any development of nuclear energy in these countries there should not exceed a few gigawatts. South Korea plans to double its nuclear capacity to around 40 GW, and Japan could either maintain its current capacity (around 40 GW) or exit nuclear energy altogether, in which case its production capacity would drop to 10 GW by 2035. Despite the uncertainties that stem from the energy policy of many countries, the worldwide growth of nuclear energy should reach between 1.3% for the low-base scenario to 3.8% for the high-base scenario. This shows the robust dynamics of the market compared to natural gas (1.7% per year on average), oil (0.5% per year) and coal (0.8% per year).

Finally, there is already a large number of new reactors being built at the moment and this will lift the overall production capacity to around 410 GW by 2020 whatever happens (NEI 2014; WEC Nuclear 2014). Nuclear energy is thus far from being in decline. Actually, there is here (like elsewhere) a shift from OECD countries to new economies. OECD countries modify their energy mix to the benefit of renewable energies, while new economies use all available technologies for electricity production to meet their ever-increasing energy needs.

3.4.3.2 Roadblocks to Nuclear Development

There are however several roadblocks to the development of nuclear energy.

Investments in uranium production are well below the ones in other industries, though they rose significantly during the past decade (Furfari 2007). The price of uranium ranged from 30 to 77 USD/kg in January 2015, after skyrocketing to up to 300 USD/kg in 2007 (Infomine 2014). More recently, the price of uranium has dropped due to the Fukushima incident and the consequences on nuclear electricity production in Japan. Nevertheless, the significant needs of China, India, and to a lesser extent South Korea will push up prices, except if Western Europe (in particular, France) and Japan decide to exit nuclear electricity.

Uranium reserves can be classified by categories of cost of production. Out of the 7.5 million tons of uranium actually identified, 9% cost less than 40 USD/kg to produce, while 23% cost between 130 and 260 USD/kg to produce (IAEA 2014) (Fig. 3.43).

The electronuclear market is not sensitive to the cost of primary materials. The cost of uranium represents indeed only 6% of the costs of production (NEA 2014). So while the cost of extraction can vary a lot and therefore influence the dynamics of the uranium market, it has only a small impact on the final electricity price. This is by far the lowest primary resource to electricity price ratio, with the exception of renewable energies (NEA 2014) (Figs. 3.44 and 3.45).

Uranium reserves represent around 130 years of production at the actual rate, and stand at around 7.5 million tons, or 88 billion tons of oil equivalent (Berkeley 2014), although these reserves are concentrated in Australia (25% of worldwide reserves), Kazakhstan and Africa (Niger, South Africa, Namibia). The ultimate potential of those reserves is not known precisely, as exploration has not been very

Fig. 3.43 Uranium reserves (IAEA 2014)

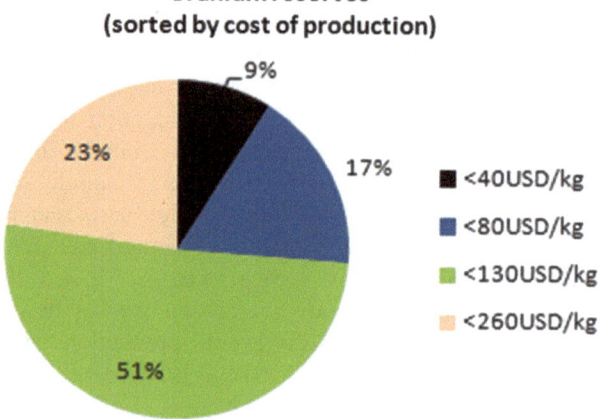

Uranium reserves
(sorted by cost of production)

- <40USD/kg
- <80USD/kg
- <130USD/kg
- <260USD/kg

Nuclear electricity cost of production

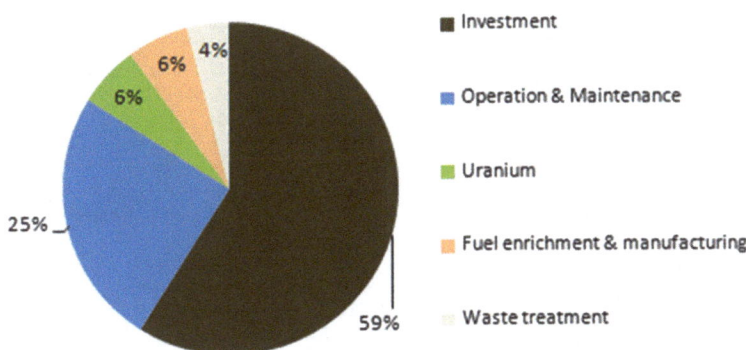

Fig. 3.44 Nuclear cost of production (NEA 2014)

Electricity production cost structure

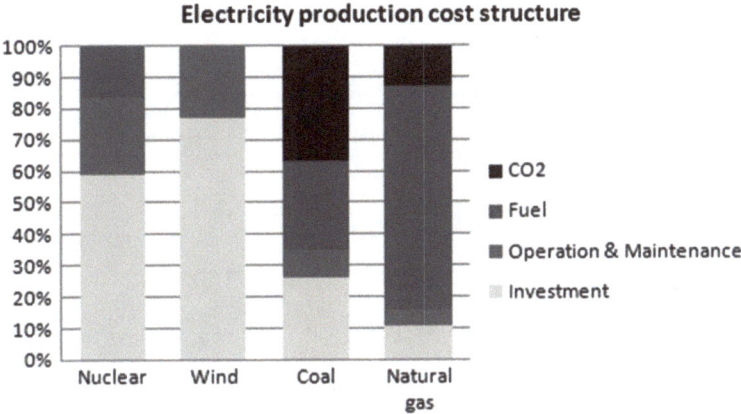

Fig. 3.45 Cost structure of electricity production (NEA 2014)

active thus far. Furfari (2007) estimates these reserves to amount up to 17 million tons, which correspond to 300 years of production at current pace.

The consumption of nuclear energy varies greatly from one region to another. More than 75% of the consumption is spread between Europe and North America. France corresponds to 17% of worldwide consumption, or 50% of European consumption (Fig. 3.46).

The average consumption per individual shows the energy choice of France, where each individual consumes around 1.4 tons of nuclear energy per year. Of electricity production in the country, 78% is indeed based on nuclear energy. Other regions of the world have more nuanced energy policies, with world average around 0.08 tons/year/individual of uranium. Theoretically, if every country were to consume as much nuclear energy on a per individual basis as France, the current

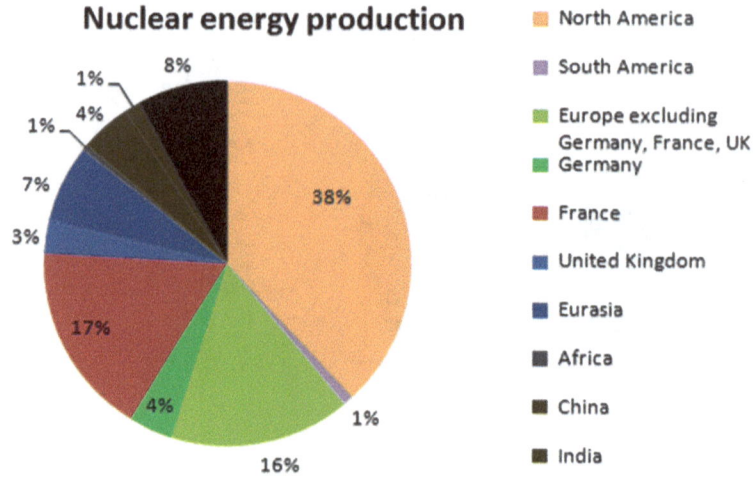

Fig. 3.46 Nuclear energy production (BP 2014)

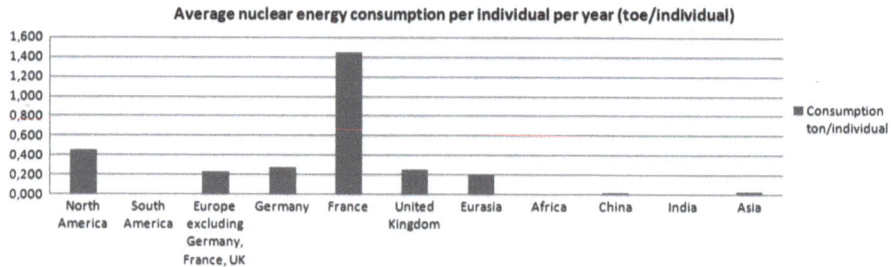

Fig. 3.47 Nuclear energy consumption per individual (BP 2014)

worldwide reserves would correspond to only 9 years of electricity production (Fig. 3.47).

The energy choices of China and India could lead to growing tensions around uranium reserves, which could be compensated by very active exploration. So the question of the actual size of the uranium reserves, which today is not a roadblock to its development, could become a challenge in the years to come.

Currently, the main roadblock to a wide deployment of nuclear energy is actually its economic profitability. Investments in nuclear power plants are long-term ones with high capital expenditures. Building a new plant may take several years, and the return on investment is extremely long. On top of these considerations, radioactive waste treatment and safety constraints are important issues which limit investments. These factors have caused the development of the nuclear industry to slow down.

Waste treatment is the primary subject of debates on nuclear energy. Most opponents to the development of nuclear power explain that countries should exit this source of energy because of the world incapacity to handle the waste. Out of all the waste generated by nuclear power plants, the "highly radioactive" one is the main source of concern. It represents around 5% of the total waste and 95% of the radioactivity. There is today no other solution than bury it deep. This waste remains dangerous for the environment during several centuries. Now, the nuclear industry is the only one to actually handle its waste. The total volume of nuclear toxic waste averages every year around 81,000 m^3 (World Nuclear 2014). This figure needs to be compared to the 300 Mm^3 of total toxic waste produced in OECD countries and for which statistics exist. Nuclear waste thus represents 0.03% of the total waste. The very toxic "highly radioactive" waste remains indeed dangerous during several centuries. Now, recycling and vitrification techniques help deal with the waste. The final stock that cannot be retreated represents around 3 m^3 per nuclear power plant and per year (World Nuclear 2014).

Finally, the security of the nuclear installations has been for decades a major concern of the populations. Efforts in this area have been important, and regulation has been further reinforced after the accident of Fukushima. The rate of incidents on nuclear plants is therefore extremely low. The Nuclear Energy Agency (2014) indicates that the non programmed interruptions of nuclear reactors have gone from 1.8 h (for 7000 h of annual operation) in 1990 to 0.5 h in 2010, which corresponds to a 75% drop in the number of incidents in 20 years. This means that there is a non programmed incident in average once every 2 years on a nuclear plant. Of course, most of these incidents have no consequences, even though three major accidents occurred in the history of nuclear electricity: Three Mile Island (1979), Tchernobyl (1986) and more recently Fukushima (2011). The worse accident was the one of Tchernobyl, which appeared to be the result of non controlled tests on the reactor, while most of the securities had been voluntarily turned off.

3.4.3.3 Summary
Nuclear energy is today limited mainly to OECD countries because of the very high capital spending it requires, as well as its ties to nuclear military applications. Currently, countries which have used nuclear energy are more inclined towards renewable energy, considering nuclear energy as an energy type of the past. For instance, Germany has decided to exit nuclear energy. Conversely, several new economies are developing or consider developing nuclear power. Nuclear electricity offers to these countries the possibility to rapidly scale up production capacity without raising at the same time the level of their dependence on fossil resources. China, India and many countries in Eastern Europe are the next frontiers of nuclear energy. Nuclear energy is therefore far from being in decline. Market growth is expected to be between 1.3 and 3.8%, more than for natural gas, and definitely much more than for coal or oil, which growth levels are not expected to exceed 1% on average per year. Electricity production should thus partially be of nuclear origin. As there are no greenhouse gases associated with nuclear electricity production, the future development of production capacity will actually contribute to the mitigation of "anthropogenic forcing".

3.4.4 Current Limits of Renewable Electricity

In the past few years, the world realized the tremendous challenge of higher consumption of electricity, as well as the waste associated with its production. Consequently, renewable energies have gained a unique position as clean substitutes for limited and polluting fossil resources.

3.4.4.1 Introduction to Various Technologies

There are four main technologies for producing electricity and heat using renewable energy sources.

Hydroelectricity is the main source of electrical power from renewable sources. For centuries already, the flow of a river has given mankind free energy for the development of a number of industrial applications. The hydro-electric dams that turn hydraulic power into electrical power mostly serve peak load requirements. They do not operate all the time and are mainly activated when a sizeable amount of energy needs to be supplied for a given period of time. The dam helps store a vital quantity of water, which can then be used when decided to rapidly supply an important amount of energy. The crucial point is to regularly reconstitute the "stock" of energy. The natural flow of the river provides this "stock". There are also pumping stations which can use the electricity provided by other plants at times where consumption is low to pump the water back upstream from the dam in order to build "stock" (Barré and Mérenne-Schoumaker 2011). This can then be reused when an additional amount of electricity is required during peak load time. When presenting hydroelectricity, one often thinks of large electric dams such as the Three Gorges Dam in China, which can produce up to 22.5 GW, or the equivalent of 20 nuclear units. There are also small hydroelectric dams which produce less than 10 MW of electricity. These plants disturb the environment to a lesser extent than large ones and are an interesting alternative for rural electrification.

The development of solar electricity has been spectacular in recent years. There are two main technologies to produce electricity out of sun power. The first one is photovoltaics. It uses sunlight to produce directly electricity from photoelectric effect in a semiconductor. This is the most deployed technology. The second is concentrated solar technology (also called solar thermal), in which solar radiation is used to heat up a coolant, similar to what happens in a thermal power plant. The vapor or gas produced is then used to activate the rotation of turbines to produce electricity via an alternator.

Wind-based electricity is also widely deployed. Wind is used to run directly the turbine which, following the same principle, produces electricity with an alternator.

Yet another energy type is geothermal energy. It recovers natural heat deep in the ground and brings it up to the surface where it is used to heat up a coolant or directly heat homes. Geothermal electricity power requires drilling more than 1500 m into the ground in order to reach high enough temperatures. When drilling is less deep, geothermal energy is essentially being used for domestic or district heating.

Finally, tidal power uses marine currents to run turbines, following the same principles as wind power.

3.4.4.2 The Spectacular but Currently Limited Development of Renewable Energy

Of the various renewable sources for generating electricity, hydroelectricity has the lion's share (Fig. 3.48).

Europe and North America lead the pack with around 21% each, followed closely by China (19%). Latin America is next in line with around 17%. The rest of the world shows limited deployment thus far (Fig. 3.49).

Latin America may rank fourth worldwide in renewable electricity generation, but 80% of its overall electricity generation is based on renewable sources, in particular, hydroelectricity. The share of renewable electricity in the global electricity generation mix in other regions generally ranges around 20%; the world average is 23% (Fig. 3.50).

The World Energy Outlook (© OECD/IEA, WEO 2012) looked at possible scenarios of how electricity generation from renewable energy sources might evolve by 2035. All scenarios converge towards a significant increase of renewable electricity generation. Scenarios from other sources also confirm this trend (Exxon Mobil 2016; Shell 2016; Greenpeace 2015). The highest increase in absolute value should occur in China, which should by then become the top renewable electricity producer in the world (Fig. 3.51).

The share of renewable energy in the total electricity mix should evolve from 24% in 2010 to more than 36% in 2035, which corresponds to a growth difference of two points versus the overall electricity generation growth (4% on average for renewable electricity growth, 2.2% on average for electricity growth worldwide). Two regions stand out: Europe would go from 28 to 50% and Africa from 20 to 42%. China would "only" evolve from 22 to 31% (Fig. 3.52).

Finally, current forecasts show that the highest contributors to the change in electricity mix are expected to be hydroelectricity and wind. Hydroelectricity is

Fig. 3.48 Renewable electricity generation per source (© OECD/IEA, WEO 2012)

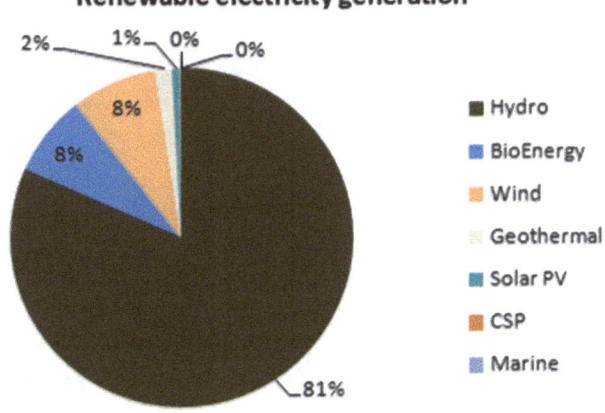

Renewable electricity generation

- Hydro
- BioEnergy
- Wind
- Geothermal
- Solar PV
- CSP
- Marine

2% 1% 0% 0% 8% 8% 81%

Renewable electricity generation

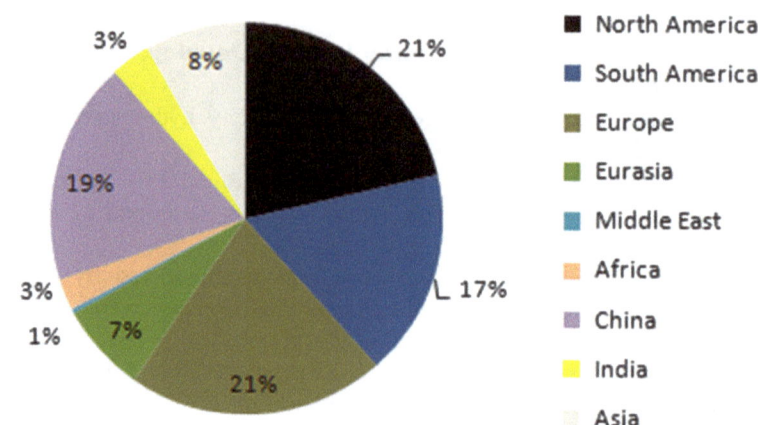

Fig. 3.49 Renewable electricity generation per region (© OECD/IEA, WEO 2012)

Share of Renewable in overall electricity generation mix

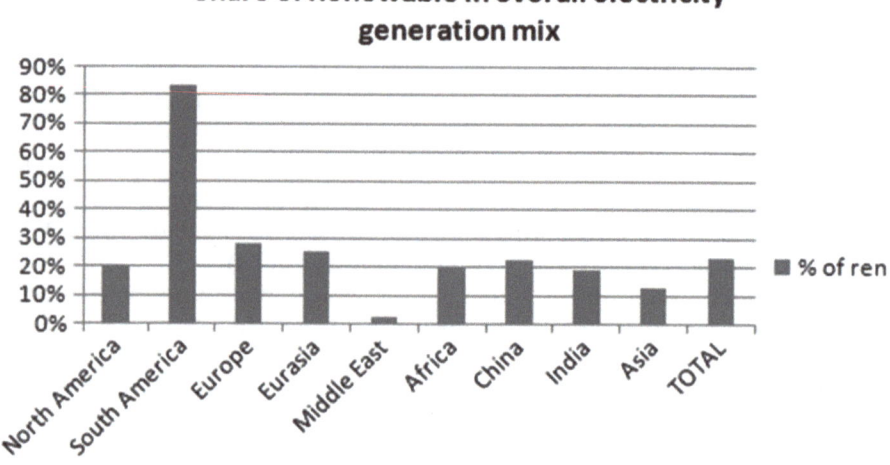

Fig. 3.50 Share of renewable electricity into overall mix (© OECD/IEA, WEO 2012)

expected to grow 2% annually, similar to electricity growth worldwide. Wind electricity, photovoltaic solar and concentrated solar power (CSP) technology are expected to grow at 8.6%, 14% and over 21%, respectively. We will see in Chap. 5 that recent developments in photovoltaic solar technology could actually yield a much higher share of solar energy in the overall mix, and prove the abovementioned forecast to be extremely conservative. Photovoltaic solar electricity could indeed be one of the major disruptions on the electricity markets in the coming years (Fig. 3.53).

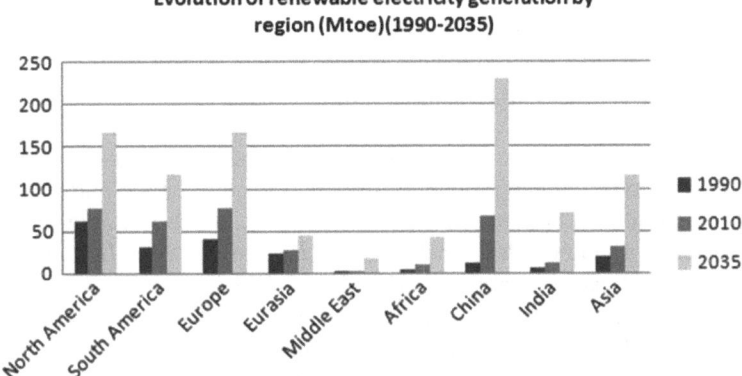

Fig. 3.51 Evolution of renewable electricity production by region (© OECD/IEA, WEO 2012)

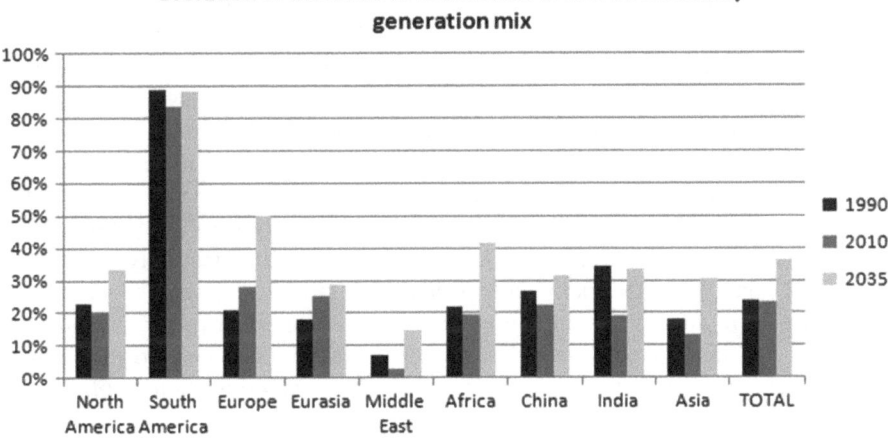

Fig. 3.52 Evolution of the share of renewable electricity into overall mix (© OECD/IEA, WEO 2012)

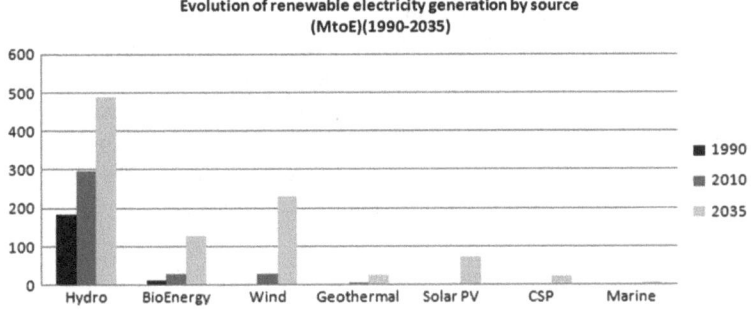

Fig. 3.53 Evolution of renewable electricity production by source (© OECD/IEA, WEO 2012)

3.4.4.3 The Main Roadblocks to Renewable Electricity Development

There are currently a number of roadblocks to renewable electricity development, despite the fact that its share of total electricity production worldwide is expected to go up to 36% by 2035 (and it is still a conservative forecast!) and that its growth is estimated at twice that of the overall growth in electricity demand.

The main issue related to renewable electricity production relates to its intermittent and not entirely reliable production. While a nuclear power plant can operate on demand and almost continuously 80% of the time, a wind farm or a solar farm generally output to the electrical network 10–20% of the time. In addition, these rates are averaged, as renewable electricity production output can vary depending on the circumstances. Electricity supply is deemed a critical utility so any disruption is considered unacceptable. Electricity demand traditionally follows a "bell" curve, with a peak load between 6 and 10 pm, and production needs to equal consumption at any one time. Since renewable electricity depends heavily upon climatic conditions, production output cannot be guaranteed and balancing production and consumption is not possible. Specific climatic conditions can lead to an excess or a deficit of production output on the electrical network which can lead to network imbalances. With the current set of technologies, renewable energies can only be a contributor to the overall electricity mix and cannot replace conventional power plants, which can be regulated to balance the network out. Hydroelectricity is the only notable exception as its contribution can be regulated, even though hydroelectricity plants operate intermittently as "stock" needs to be rebuilt on a regular basis. Outside of hydroelectricity, the current penetration rate of renewable energy is around 5% in average. The penetration rate may vary depending on the particular situation of the networks as well as the specific electricity production mix. In Denmark or Germany, the penetration ratio is higher, the effects of variability of production being compensated by coal thermal power plants (very polluting) and interconnections to the European transmission network. In France, the rate remains low as 78% of the current electricity production is of nuclear origin. Nuclear power is indeed very stable but not flexible. The share of nuclear power production thus prevents the overall system from accommodating more than a certain volume of intermittent, as it cannot provide the flexibility required (Durand 2007). Most experts think this rate can easily increase around 10–15% without significant impact on the network. Above this level of penetration, the penetration of renewable energies leads to important restructuring of conventional generation and to the setup of necessary flexibility mechanisms. It is generally admitted that the network stability can still be managed up to 40% of renewable penetration, but higher ratios are today unknown. The current growth of renewable production output therefore raises many questions and issues on how to manage the electrical network. Possible disruptions are however on the verge to occur. They could reshape completely the electricity market landscape. They will be looked at in the Chap. 5.

3.4.4.4 Other Perspectives for Renewable Energy

Besides electricity production, renewable energy can be used to produce heat. Actually, heat generation was historically the first use of renewable energy—biomass (wood, waste, etc.) is the oldest way of heating up a space.

According to the International Energy Agency (2012), about 1000 Mtoe of biomass was used for heat generation in 2010. The vast majority of it (77%) was related to a traditional use of energy (individual heating, lighting, cooking).

Renewable energy is very interesting for heat generation, whether it is space or water heating. Heat can be stored to a certain extent, making renewable energy (geothermal, solar) very suitable for heating systems. The use of renewable energy for heat production could thus have a significant impact on fossil resources consumption (© OECD/IEA, Heating 2014).

3.4.5 Electricity Market Challenges

One main characteristic of the electricity market is that electrical energy is not stored. It is thus mandatory to permanently balance production and consumption. The lack of balance has devastating consequences for the electrical network, including frequency and voltage variations. Beyond certain limits, these variations can yield damage to connected equipment as well as blackouts. Stable and good-quality electricity supply is thus essential.

At the end of the Second World War, electricity production, transportation and distribution infrastructures were built by giant vertical companies. These companies were publicly owned and strongly regulated as electricity supply was considered a matter of national security, and strategic for the economic rebound of economies exhausted by the war. Electricity distribution was stable, vertically integrated, and predictable. The stability of the electrical network was primarily ensured by the simplicity of the distribution. Large power plants (bulk power) were interconnected on a transmission network. This network then supplied energy to a number of distribution networks, which then supplied energy to the millions of final consumer points. The stability of the transmission network was to be maintained at all times and the distribution networks had few constraints. The other important factor of stability was the predictability of consumption. The consumption remained fairly predictable from 1 h to another and from one season to another. As a result, all sources of energy were mobilized and organized centrally to supply the required energy at a given time for a given day. Large companies from the sector ensured at all times the overall stability of the network, and the maintenance of the infrastructure.

Two factors combined to disrupt the current system on which electricity markets today operate (© OECD/IEA, Power Transition 2012). First, the vertically integrated public companies involved with electricity production, transportation and distribution were more and more decoupled as independent companies. The motive was to deregulate the market in order to reduce costs and reach a higher level of efficiency. Second, a massive amount of subsidized renewable energies was introduced on the network, with the objective to substitute eventually thermal power plants. The second was partially enabled by the first one as the progressive deregulation of the power generation market allowed for private investors to kick in and to benefit from specific market arrangements related to renewable energies, developed by market regulators. With the drop of costs related to volume, renewable energies now become progressively competitive in various countries and can further penetrate the power generation market without subsidies in place.

These two changes deeply destabilized the electricity markets. In Germany, a transmission operator had to run more than a thousand operations to rebalance the network in 2010 (© OECD/IEA, Network Infrastructures 2013), compared to two on average per year before these changes. In the United Kingdom, forecasting errors skyrocketed from 7% before the introduction of renewable energy to more than 28% (© OECD/IEA, Network Infrastructures 2013).

The increasing penetration of renewable energies has yielded a missing opportunity for conventional generation and therefore economic impact. Worse, as the volume of renewable energies grows and becomes naturally competitive with conventional generation, many conventional power plants face now situations where their profitability is threatened. Several times in the recent years, the daily "spot" price in Germany became negative as the renewable farms produced in excess and it was more rational for independent producers to supply even at zero (or negative) selling price than to stop their power plants (because of the cost to turn off and on again). This situation obviously cannot last. This accelerated penetration of renewable energies is thus leading to a heavy restructuring of the conventional generation market.

Additionally, the lack of predictability of electrical consumption and of energy balance over the network due to renewable energies presents a number of issues. First, localized overloads over the transmission or distribution network lead to congestion. This may lead to supply disruptions in certain parts of the network. Voltage imbalance in some nodes of the network can also occur, due to the lack of coordination between different producers. As well, unplanned incidents may lead to an imbalance between supply and demand in certain parts of the network. Finally, most renewable energies are connected to the distribution network (unlike bulk power which is connected directly on the transmission network), and therefore create energy flows in all directions (downward to consumers, upward towards the transmission network), which lead to new issues of imbalance and localized congestion. As a consequence, the volume of reserves and services required to balance out the network and secure the stability of the grid goes up, and so does its cost for final consumers.

Deregulation has as well created an increased complexity, since it led to new behaviors which are not always efficiently coordinated (© OECD/IEA, Power Transition 2012). The vertically integrated public companies used to coordinate all decisions in order to ensure optimal operation of the network, both from a stability and an economical standpoint. The decoupling of functions changed things. In countries where electric utilities are not verticalized, the electricity market is indeed operating under the principle of merit order within a market "pool", also called "wholesale" market. Within the "pool", each independent producer submits the volume it intends to supply and its proposed selling price 1 day ahead. The market operator ranks the different offers by price and the less expensive producers are selected to be part of the electricity generation pool (often renewable energies are also prioritized by default). Once this selection is done, the operator simulates the impact of this set of generation capacities on the transmission and distribution network and identifies existing congestion or imbalance problems. If network stability is threatened, the operator modifies the list of selected producers. It also makes sure that a sufficient volume of production reserves will be available to preserve network stability should unfavorable incidents occur. Services markets have thus been established for the network operator to purchase a variety of balancing and ancillary services, required to balance in real-time load and supply and maintain network stability ahead of possible incidents. Finally, a retail market also exists, for final consumers, who do not have access to the wholesale market. The price of electricity on the retail market is a result on the price traded on the wholesale market as well as the various services and the cost of transmission and distribution of energy down to the final consumer point. With these changes, most players are typically incentivized on the profit they can generate, not anymore on their contribution to the network's overall balance, which turns out to be a specific cost charged separately. The importance of the market regulator and operator has thus become paramount.

Market regulators need to find solutions to these complex issues, while encouraging the further deployment of renewable energies.

The first step to solving this complex topic is to properly measure what goes on the grid in real-time. The first electricity networks were designed more than 50 years ago. At that time, the consumption profile was predictable and energy was distributed in a single direction. As everything was designed to operate properly at the nominal level, there were no measurement instruments on the network. Everything has changed since then. Energy transits in all directions and it becomes very difficult to predict accurately production and consumption profiles from different sources or loads. It then becomes necessary to control in real time what is happening on the electrical network. On transmission networks, Phasor Measurement Units (PMUs) and dynamic line rating help operators understand in real time if the network is imbalanced. Most distribution operators are only starting to install instrumentation on their networks. The needs move from high voltage down to low voltage and then the consumer point. These new points of measurement allow better control over complex networks which not only let energy "pass through", but also deal with multiple interconnections. So distribution networks might become as complicated to handle as transmission networks. Measurement at all levels, distribution management systems and local automation in substations help operators to localize and manage incidents faster, as well as better understand network balance issues while optimizing load transportation.

The second step is to design a more efficient and more transparent market for the commodity. Usually, prices are fixed on day ahead in the wholesale market, at the time of trade closing. The uncertainty of power production capacity intra-day (in particular from renewable) is thus managed on other markets, in particular the various services' markets, which basically have to deal with the overall inefficiency of the system. Some market operators are already working at extending the duration of trade up until 1 h before the actual operation. Postponing the closing of trade would allow the operator to minimize forecasting errors and therefore the volume of services required to be purchased. The available generation capacity becomes indeed more reliable as we get closer to the time of actual operation. As well, in some regions of the world (in particular, the United States), market operators have already introduced a nodal marginal pricing system, which integrates in the price, for a given producer, the constraints its production imposes on the network depending on its location, notably in terms of congestion. This pricing system can be applied to each network node or to a region; the smaller the region to which a single price applies, the more the transparency. Increased granularity in both time and geographic localization on the market is expected to make it more efficient and more transparent. In parallel, many countries start to create reserve capacity markets. These markets consist of paying power plants for them to be ready in case of untoward incidents so that they can contribute on request a given volume in a given time frame. These capacity markets aim at ensuring a proper revenue for conventional generation units which are essential to overall network balance and stability, and which would otherwise have to be decommissioned.

The third step towards solving this complex issue is to build in the overall system operation a higher level of flexibility. A dominant opinion is that energy producers and final consumers are not sensitized enough to the impact of their action on the overall network operation, and therefore are unable to take any decision to limit it. The "Smart Grid" revolution intends to involve them in network operation. At the renewable operator level, market regulators are working on grid codes to make the integration of renewable energies on the grid more stringent. The consequence is that it becomes more complicated for renewable operators to drive properly their return on investment. As a side issue, wholesale prices of electricity tend naturally to be the most impacted when the share of renewable is the highest on the network. Consequently, the more renewable energies on the network, the lower the benefits of renewable operators on the wholesale market. This is leading renewable operators to aggregate vast quantities of renewable production units (and sometimes demand) in order to get a higher level of flexibility in their overall output, thus market participation, and possibly enable them to also play on different markets such as the services' markets. This is what lies behind the emerging concept of "virtual power plants". The other element of the equation is final consumers. The installation of smart meters, capable to monitor in real time the consumption and to transmit the information to a central system, offers remarkable perspectives. From the information collected by the smart meters, it becomes possible to understand the consumption profile of millions of users and to connect with them. The small user may serve the overall network stability by better managing its energy consumption. The main challenge here is the reduction of the "peak load", which is a sudden rise of the electricity consumption which generally happens every day between 6 and 10 pm. This is the time where the consumption is the highest. It can reach then up to two times the lowest consumption level of the day. Now, the network and the production capacities are dimensioned to meet the "peak load" level. Controlling and decreasing the volume of this peak by a better collective usage of electricity (more flexible) would have considerable consequences on the production as well as on the overall economy of the sector. The emergence of demand side management programs and market aggregators (consolidating various consumers) are first signs of an increased flexibility at the demand level.

As a summary, the electricity markets are undergoing a significant evolution, essentially due to the deregulation and the penetration of massive amounts of variable renewable energies. This trend will not be slowed down since renewable energies become more and more competitive with regards to conventional generation. Consequently, the market is facing a disruption. The necessary restructuring of the conventional markets will however not be enough. The new electricity network will represent a complete change of paradigm as compared to the historical one. Digital technologies will be paramount into enabling this transition. First, the network needs to be digitized, with measurement systems and sensors at all levels of the grid, in order to enable real-time management. Then, the electricity markets' structure needs to be revisited to enable a higher level of granularity and precision, as well as protecting a certain amount of traditional capacity. Finally, the key word behind "smart grids" is certainly flexibility. Flexibility will be at the core of the electricity organization of the future. Flexibility at the production level will minimize the impacts of the lack of predictability and the services to be procured to face incidents. Flexibility at the consumption level will help adjust in real-time the load curve to meet the overall network constraints, in particular at times of peak consumption.

3.4.6   The Complexity of Electricity Price Calculation

Electricity price is a complex variable which may vary strongly across regions and users (Fig. 3.54).

It depends both on the constraints that apply to local electricity production (cost of production, cost of operation and maintenance of the transmission and distribution networks) and on the pricing policy decided by regulators (taxes to subsidize certain activities, such as renewable for instance). In general, electricity price is set to cover the various costs borne by producers and network operators.

On average the cost of production represents 50% of the total electricity price, the cost of the network 25%, and the various taxes and subsidizes the remaining 25% (© OECD/IEA, WEO 2012).

Electricity production cost is made of a number of elements which add up to make the price. Of these elements, the most important are generally the fixed costs of the plant and the cost of fuel. Variable costs and some taxes (such as carbon tax applied in the European Union) contribute to these two costs but are much lower (© OECD/IEA, WEO 2012). The structure of costs of electricity production however varies strongly from one technology to another. Nuclear power and

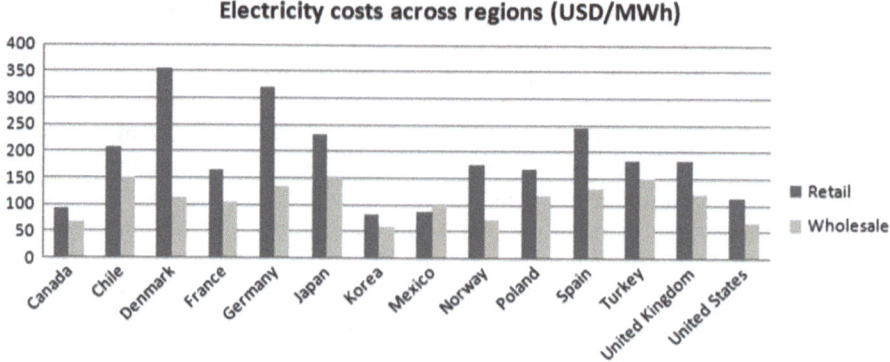

Fig. 3.54 Electricity costs across regions (© OECD/IEA, Electricity 2012)

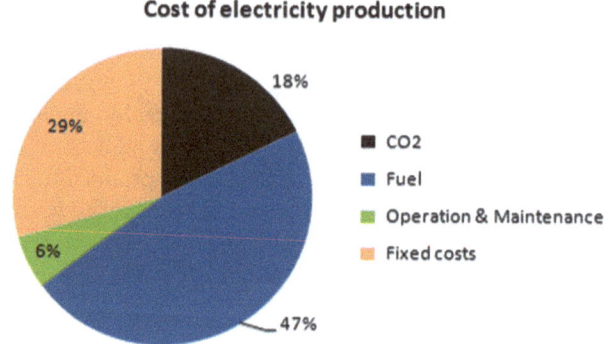

Fig. 3.55 Example of cost of electricity production (© OECD/IEA, Electricity 2012; © OECD/IEA, WEO 2012)

renewable energies are traditionally more dependent on fixed costs, while natural gas and coal power plants are more sensitive to fuel costs and variable costs (Fig. 3.55).

Network cost is very complicated for the consumer to understand. Figure 3.56 (© OECD/IEA, Networks Infrastructures 2013) shows the operational costs of some transmission operators in Germany. This corresponds on average to 65% of total network costs, which also include capital costs, depreciation and maintenance costs. It shows the very high share of operational reserve payments (reserve capacity) and the costs for network losses which need to be incurred by network operators. These costs are very significant for network operators.

In the end, production costs vary a lot depending on the type of production capacity. Network costs also vary depending on the volume of electricity shared, and the conditions of operations. Some disruptions could happen in the coming years with regards to electricity costs, in particular for retail electricity. They will be looked at in Chap. 5.

Fig. 3.56 Cost of the electric network (© OECD/IEA, Electricity 2012; © OECD/IEA, WEO 2012)

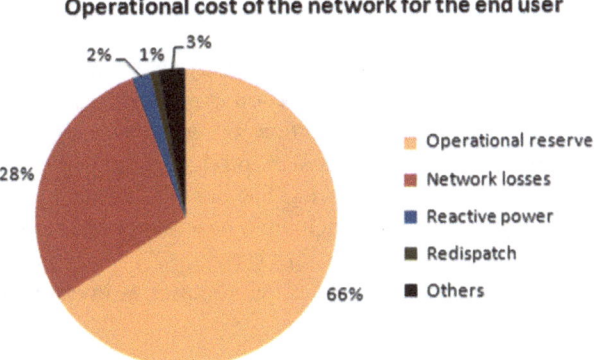

Operational cost of the network for the end user

- Operational reserve
- Network losses
- Reactive power
- Redispatch
- Others

2% 1% 3%

28%

66%

3.4.7   Summary

Electricity production will grow by more than 70% in the next 20 years, driven by world population growth and, more importantly, by the improvement of living standards in new economies. Indeed, it is increased urbanization and consumption of goods which primarily cause the surge in consumption. The impact of economic development in new economies will have considerable consequences on electricity production. Production growth will essentially take place in Asia; the economic development of China in particular should pull global growth up. Beyond 2035, electricity consumption should continue to increase, driven this time by other regions of the world such as India, the rest of Asia and Africa as they achieve their own economic transition. Electricity production is mainly achieved using fossil fuels. The process of producing electricity is extremely inefficient, and consequently the increase in production will have extreme consequences on the consumption of primary fossil resources, and on greenhouse gas emissions. With an average of 30–40% of yield, around two thirds of the energy consumed in a conventional power plant is wasted. Beyond the waste, the need for primary resources to produce electricity can also create in certain regions an unbearable dependency on fossil fuels. The tension on primary resources shall thus continue to accentuate. The massive introduction of renewable energy helps solve these two issues. They have indeed no impact on the environment and they help reduce the dependency on fossil fuels. However, the introduction of renewable energy is complex. Market deregulation and the integration of intermittent sources of renewable energy have made electrical network management considerably more complicated. Electricity cannot be stored therefore the production and the consumption must be balanced out everywhere and in real time. Intermittent sources of energy have therefore a strong impact on this balance. The complexity of this change lies behind the "Smart Grid", which is a major challenge for the industry as well as a fantastic opportunity to reinvent 70-year-old rules of electrical grid management.

3.5   Massive Needs for Investments in New Capacities

The overall growth of primary energy consumption in the coming 20 years requires that new capacities of production as well as the infrastructure to distribute it be put in place. Electricity production will grow very significantly and will also require a number of new capacities of production and the renovation of very large and disperse electrical networks. The demand for energy will then drive a considerable flow of investments, which are a necessary condition for sustained development of the global economy.

This chapter summarizes the volumes of investment by region and by sector.

3.5.1   Global Overview

The cumulated amount of investments between 2014 and 2035 in the energy sector as a whole is estimated at up to 48 trillion dollars (© OECD/IEA, Investment 2014). This corresponds to an increase of 30% versus the average in the last 10 years. Half of these investments correspond to the extraction, transportation and transformation of primary resources. These are followed by investments associated with the production and the distribution of electricity, and those aimed at improving energy efficiency (Fig. 3.57).

New economies represent the two thirds of the targeted geographic areas of investments. Asia represents one third of the total, and China alone 15% of the total (Fig. 3.58).

Investments in energy will vary strongly across regions. North America will lead the pack, investing as much as 6 trillion dollars in the primary resources sector, which corresponds to 27% of worldwide investments. Just over a tenth of worldwide investments in primary energies will be made in the Middle East, confirming the geopolitical shift in primary energy. In the electricity sector, investments will be made everywhere, in particular, China, which should represent 21% of worldwide investments. For energy efficiency investments, Europe should lead with a 28% share of worldwide investments, followed by North America and China (Fig. 3.59).

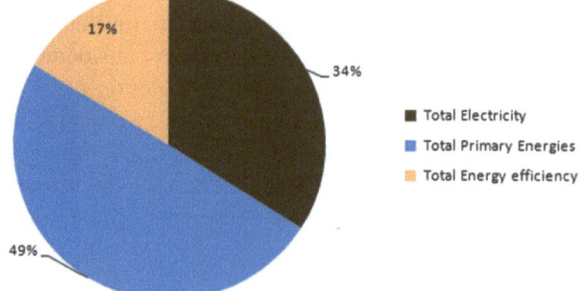

Fig. 3.57 Worldwide energy investments (© OECD/IEA, Investment 2014)

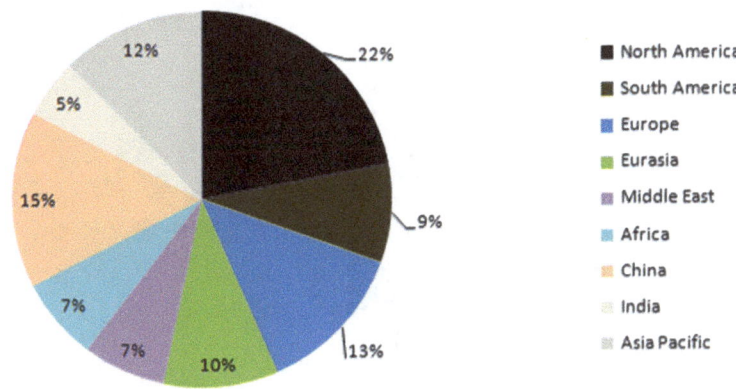

Fig. 3.58   Energy investments per region (© OECD/IEA, Investment 2014)

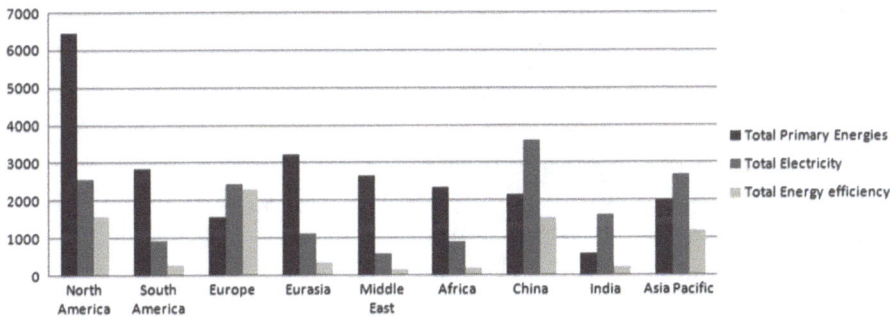

Fig. 3.59   Energy investments per region and sector (© OECD/IEA, Investment 2014)

3.5.2   Primary Energy Resources Investments

The primary energy market should attract investments worth 24 trillion dollars till 2035. Of these, 58% will go to oil and 37% to natural gas. Investments in coal mines should not exceed 4% of the total investments (Fig. 3.60).

These investments are expected to be equally spread across the regions, with the exception of North America, which should be investing massively in unconventional oil and gas. The American continent (North and South America) should represent 40% of the worldwide investments, with North America accounting for 27% (Fig. 3.61).

Worldwide investments in the primary energy sector

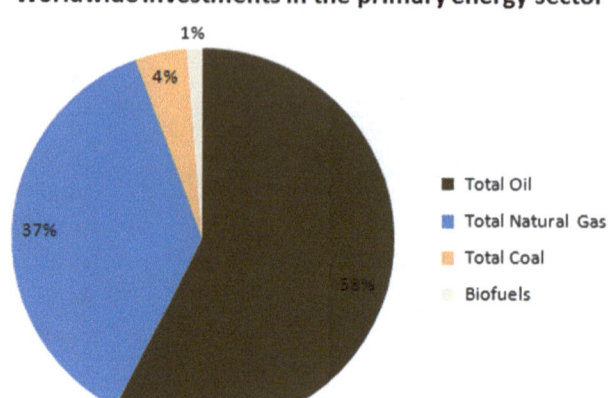

Fig. 3.60 Primary energy investments (© OECD/IEA, Investment 2014)

Worldwide investments in the primary energy sector

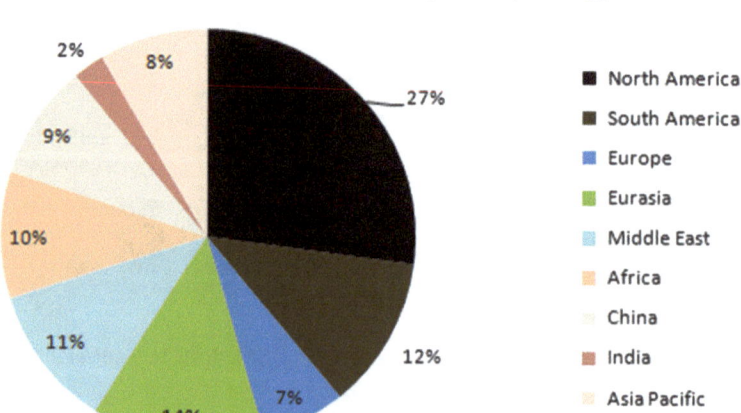

Fig. 3.61 Primary energy investments per region (© OECD/IEA, Investment 2014)

In the oil market, the upstream segment (exploration and production) should dominate with an estimated 80% of total investments. Asia is expected to reinforce significantly its transportation and distribution capacities in order to sustain both production and demand growth, as well as to diversify the origin of resources. North America will gain a significant share of the upstream investments. In South America, Brazil and Venezuela will be the primary targets for investments (Fig. 3.62).

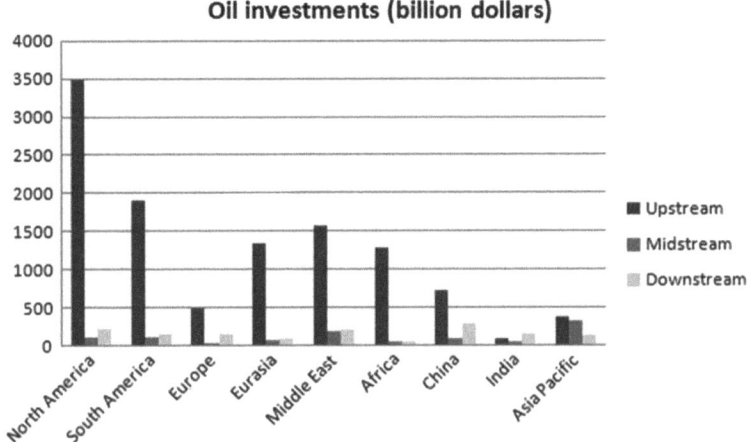

Fig. 3.62 Oil investments per region (© OECD/IEA, Investment 2014)

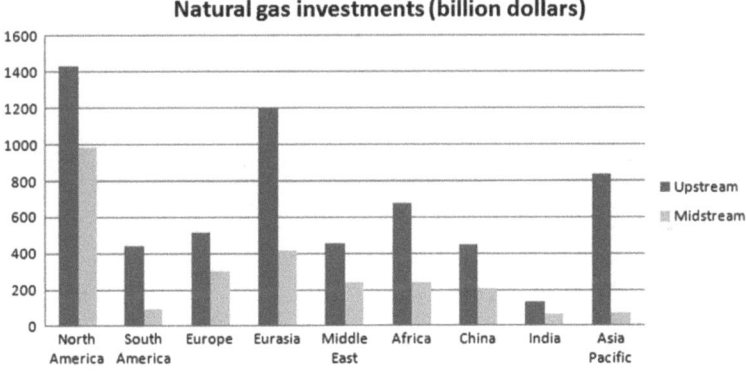

Fig. 3.63 Natural gas investments per region (© OECD/IEA, Investment 2014)

In the natural gas market, North America and Eurasia will dominate. Asia Pacific (including Australia) will see a number of important investments as well. In this geography, investments in transportation (midstream) are more important than in the oil market, due to the inherent issues related to natural gas transportation. North America invests also considerably in transportation and distribution infrastructures to supply its shale gas (Fig. 3.63).

The coal market will continue to be dominated by Asia. The International Energy Agency (2014) expects China to keep investing in coal mines to meet the growing demand for electricity in the country. Other Asian countries should follow suit, although at a slower pace (Fig. 3.64).

Finally, biofuels should continue to develop, essentially on the American continent, and notably in South America, where they are already fairly well developed. Asia should remain massively dependent on oil (Fig. 3.65).

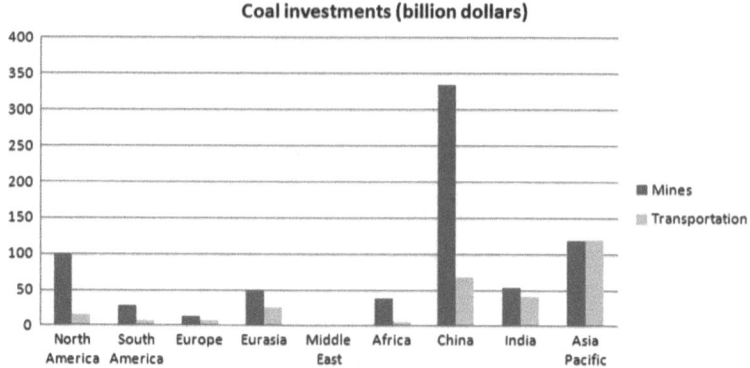

Fig. 3.64 Coal investments per region (© OECD/IEA, Investment 2014)

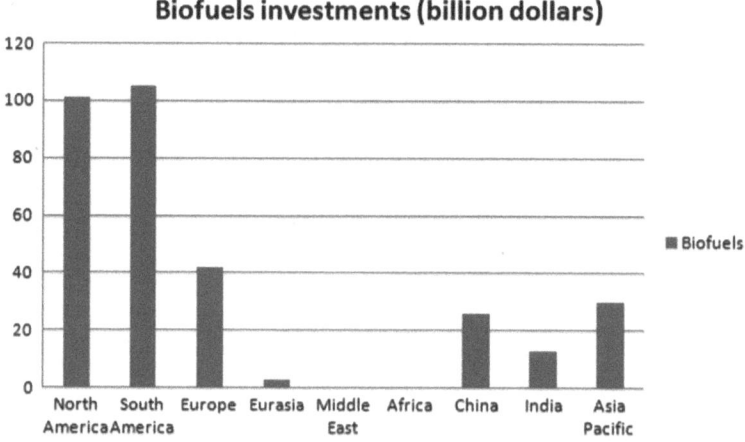

Fig. 3.65 Biofuels investments per region (© OECD/IEA, Investment 2014)

3.5.3 Electricity Market Investments

Total investments in the electricity market should reach 16 trillion dollars by 2035. This estimate from the International Energy Agency (2012) is slightly lower than other sources such as Greenpeace (2015) or the World Energy Council (2013) which estimate these investments to reach up and beyond 20 trillion dollars by 2035. The majority of these investments will go towards building new production capacities, although transportation and distribution should represent around 7 trillion dollars of investments. Renewable energy will take a two thirds share of investments in the production sector (Fig. 3.66).

The OECD countries should represent one third of the investments by 2035. These investments mostly correspond to the renovation of ageing electrical networks and to the integration of renewable energies which will replace partially

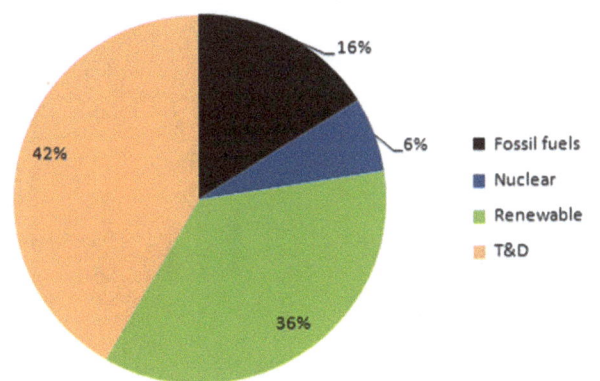

Fig. 3.66   Worldwide electricity investments (© OECD/IEA, Investment 2014)

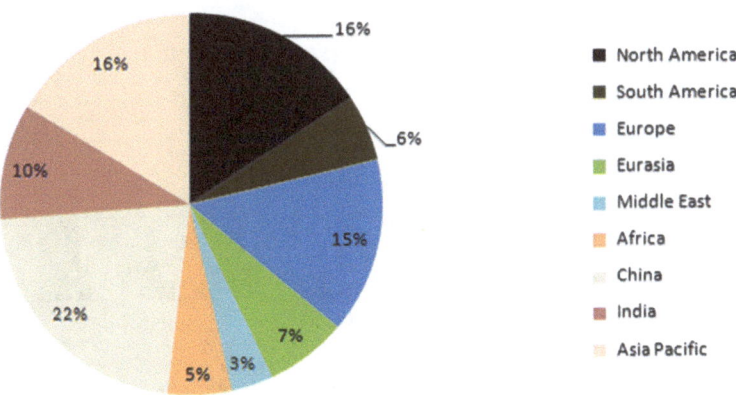

Fig. 3.67   Electricity investments per region (© OECD/IEA, Investment 2014)

conventional power plants. Asia (in particular, China) is expected to account for half of the investments worldwide. As already explained, the economic transition in the region will drive very strong growth of the demand for electricity, which in turn will create a need for considerable investments (Fig. 3.67).

Renewable energies would essentially grow in Europe, in Asia and in North America. Nuclear energy shall continue to grow in China and in Europe. Asia should continue to invest in conventional thermal power plants (based on fossil fuels) to meet growing demand. Likewise North America, which holds important reserves of natural gas (Fig. 3.68).

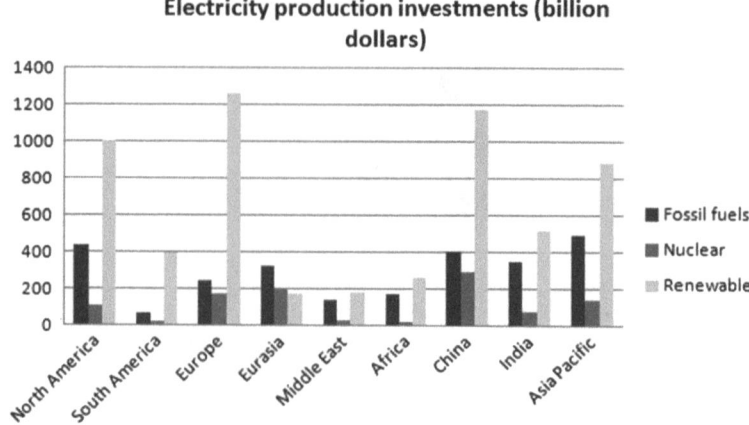

Fig. 3.68 Electricity investments per region and source (© OECD/IEA, Investment 2014)

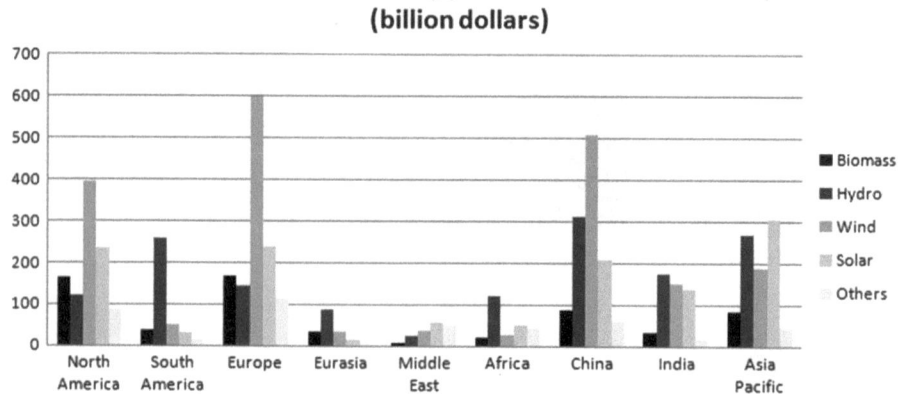

Fig. 3.69 Renewable electricity investments per region and source (© OECD/IEA, Investment 2014)

The renewable energy market is expected to be dominated by wind energy, which should correspond to one third of the total investment, followed by hydroelectricity and solar generation (around 20–25% each).

The northern hemisphere would mainly use wind energy, while hydroelectricity would dominate in South America, Russia, South Africa and India. We will see in Chap. 5 that these forecasts could evolve significantly in light of the current technical disruptions in the renewable energy market (Fig. 3.69).

Finally, the investments in transmission and distribution of electricity are expected to amount to 7 trillion dollars by 2035, in tandem with the renovation of electrical networks. Distribution should represent 75% of those investments. Indeed, the main challenge is in distribution. Distribution networks need to be

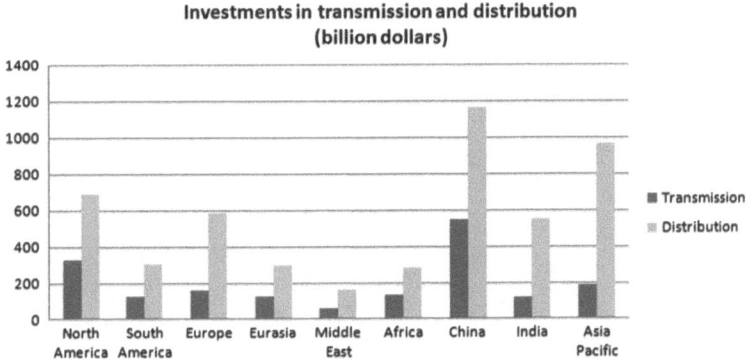

Fig. 3.70 Transmission and Distribution investments (© OECD/IEA, Investment 2014)

dimensioned and equipped to sustain and manage intermittent flows of renewable energies flowing in all directions. Asia should represent 50% of the investments in the domain, with China taking a sizeable share (Fig. 3.70).

3.5.4 The Energy Efficiency Market

New investments in energy efficiency should total up to 8 trillion dollars by 2035, according to the International Energy Agency (2014). This figure is confirmed by The New Climate Economy (2016). The International Renewable Energy Agency comes up with a slightly lower figure (Cleantechnica 2016). A 62% share of these investments is expected to be essentially realized in the transportation sector, even though it only represents 28% of final energy consumption. The industry sector, which represents 33% of final energy consumption, would only receive 9% of the projected energy efficiency investments (Fig. 3.71).

These investments shall mainly be made in OECD countries, particularly those in Europe, which represents 30% of worldwide investments. Asia should attract one third of the investments; it now has a 40% share of energy demand and should account for two thirds of energy demand growth in the next two decades. According to the International Energy Agency (2014) energy efficiency investments in other regions will be very limited. The energy efficiency market thus appears to concern today primarily countries with a large energy demand footprint (Fig. 3.72).

As mentioned above, investments in the industry sector are expected to be very small, although it is the primary source of greenhouse gas emissions (with electricity production). Despite this, industry sector investments are well balanced between energy-intensive industries and others. They shall mainly take place in Asia, particularly China, which corresponds well to the global industrial footprint (Fig. 3.73).

Worldwide investments in the energy efficiency sector

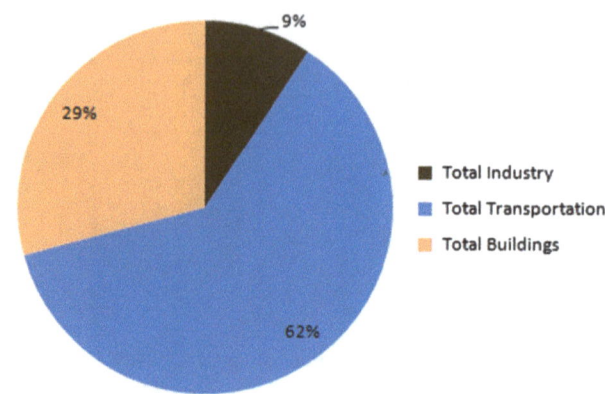

Fig. 3.71 Worldwide energy efficiency investments (© OECD/IEA, Investment 2014)

Worldwide investments in the energy efficiency sector

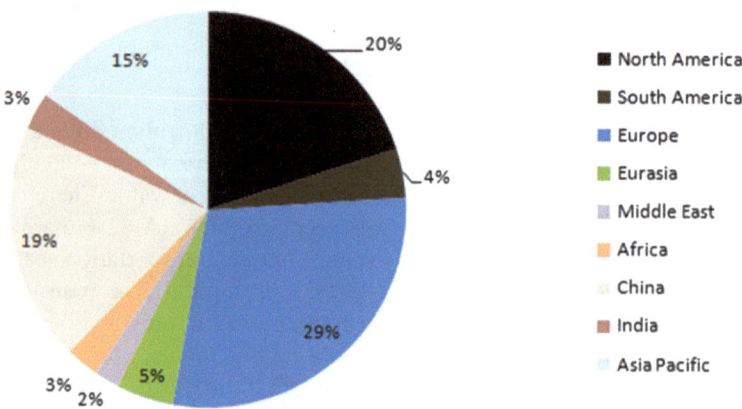

Fig. 3.72 Energy efficiency investments per region (© OECD/IEA, Investment 2014)

Energy efficiency investments in buildings are expected to mainly be in Europe, and to a lesser extent North America and Asia (Japan, Australia, Korea). Investments are not foreseen in the rest of the world (Fig. 3.74).

Of the future energy efficiency investments in the transportation sector, most of those in Asia would be in marine and air infrastructure, pacing the rise of mobility in the region. In other regions around the world, road transportation optimization shall dominate, mainly in Europe, China, and North America (Fig. 3.75).

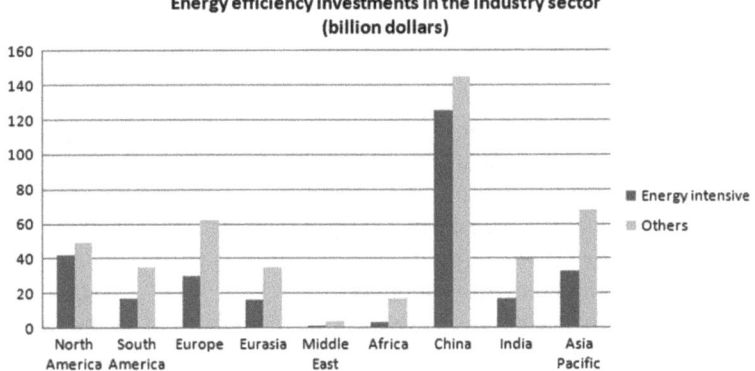

Fig. 3.73 Energy efficiency investments in industry (© OECD/IEA, Investment 2014)

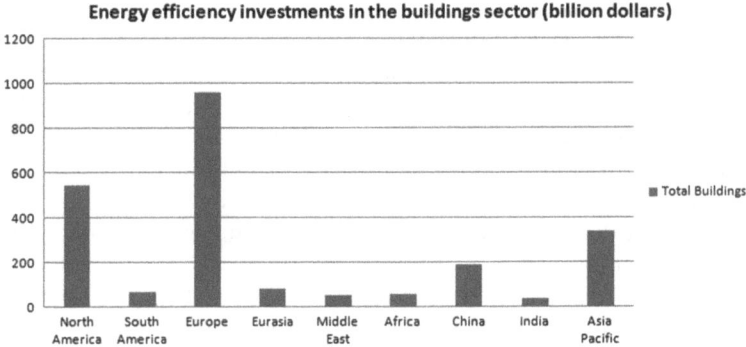

Fig. 3.74 Energy efficiency investments in buildings (© OECD/IEA, Investment 2014)

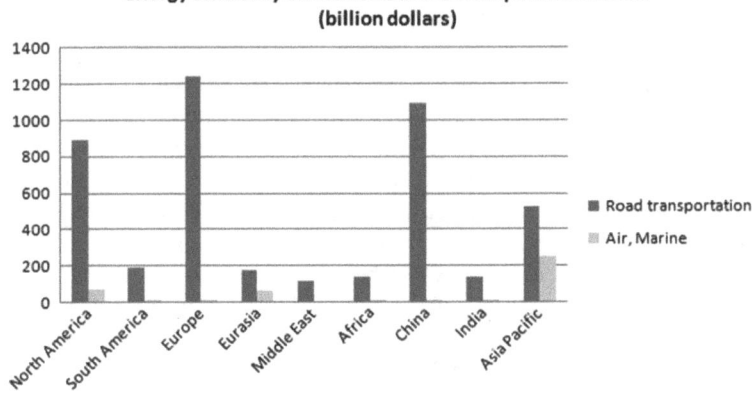

Fig. 3.75 Energy efficiency investments in transportation (© OECD/IEA, Investment 2014)

3.5.5 Summary

The cumulated volume of investments in the energy sector worldwide by 2035 would reach 48 trillion dollars, an increase of 30% compared to the average of the last 10 years (© OECD/IEA, Investment 2014). More than 2.2 trillion dollars would be invested yearly, compared to 1.6 trillion invested on average in the last 10 years. The increase in investments confirms that the energy transition is in progress and that new economies are catching up with mature ones. It also confirms while the world needs more energy going forward, it is trying to limit its dependency on fossil fuels and its energy footprint.

Huge disparities shall continue to exist in the domain: half of the investments would be in the primary energy sector, one third in the production and distribution of electricity, and the rest in energy efficiency.

The primary energy sector would be dominated by the American continent, in particular, North America, with oil and gas attracting the vast majority of investments (95% of the total). The coal market will remain dominated by China and, to a lesser extent, the rest of Asia.

The electricity sector would invest massively in the renovation of transportation and distribution infrastructures (42% of total investments), in particular, the distribution networks. New capacities of electricity production would mainly be based on renewable energies (36% of total investment, two thirds of the investment in new production capacities). They would mainly be made in Asia, Europe and North America. In Asia, these investments shall support the economic transition of the region, while in Europe and in North America, they would correspond to a change in energy mix, linked to the decommissioning of end-of-life thermal power plants. Wind is expected to be the dominant renewable energy source, followed by hydroelectricity and solar.

Finally, the emerging energy efficiency sector would represent 17% of total investments. Of these investments, 50% would go to automobiles and urban transportation, while investments in buildings and in industry would only represent 3 trillion dollars by 2035, or less than 170 billion dollars per year. The world is thus still far away from a global energy efficiency policy. The picture is different across the world's geographies. Energy efficiency investments are expected to be significant in Europe, where they shall represent 36% of the total volume of energy investments. The corresponding percentage for the Middle East is however less than 5%.

According to the White House (2014), the absence of an active mitigation plan for climate change could result in a sustained 0.9–2% slowdown of worldwide GDP by 2050. Obviously, the negative effects on climate change are not yet fully visible and some of the costs incurred may not be recorded today as a result of political inaction. This is why the current impact on global GDP does not necessarily justify everywhere the implementation of energy efficiency actions. The slowdown would obviously cumulate year after year. Indeed, 2% less growth in global GDP per year represents more than half of the total growth, and therefore would cumulate in forty years to up to a 50% loss in absolute value. Seen from that angle, the return on investment on energy efficiency is unchallengeable. The earlier the investments are made, the higher the results over a given period of time.

References

Advisor Perspectives (2015) http://www.advisorperspectives.com/articles/2015/03/17/mlps-will-weather-the-storm

Barré B, Mérenne-Schoumaker B (2011) Atlas des énergies mondiales. Editions Autrement, Paris

Berkeley (2014) http://astro.berkeley.edu/~wright/fuel_energy.html

BNP (2015) Investors' Corner. http://investors-corner.bnpparibas-ip.com/markets-strategy/crude-oil-market-briefing-lower-longer/

BP (2009) http://www.jaimelafinance.com/2011/02/petrole-quels-sont-les-flux-mondiaux.html

BP (2014) BP statistical review (on website)

Carto (2014) Le Monde en Cartes, Mars-Avril 2014. www.carto-presses.com

Cleantechnica (2016) Global energy efficiency investment will = $5.8 trillion by 2030. http://cleantechnica.com/2016/02/05/global-energy-efficiency-investment-will-5-8-trillion-by-2030/

Conca J (2015) In Forbes, July 2015. http://www.forbes.com/sites/jamesconca/2015/07/22/u-s-winning-oil-war-against-saudi-arabia/

Desjardins (2007) http://www.desjardins.com/en/a_propos/etudes_economiques/previsions/en_perspective/pe_0710a.pdf

Durand B (2007) Energie & Environnement, Les risques et les enjeux d'une crise annoncée. Collection Grenoble Sciences, EDP, Les Ulis

Economic Theories (2008) http://www.economictheories.org/2008/11/malthus-and-ricardian-theory-of-rent.html

EIA (2013) Technically Recoverable Shale Oil and Shale Gas Resources: an assessment of 137 shale formations in 41 countries outside the United States. (Energy Information Agency). http://www.eia.gov/analysis/studies/worldshalegas/

EIA (2015) Energy Information Agency. http://www.eia.gov/forecasts/steo/report/global_oil.cfm

Energy Matters (2014) http://euanmearns.com/oil-price-wars-who-blinks-first/

European Central Bank (2004) Jimenez-Rodriguez R, Sanchez M. https://www.ecb.europa.eu/pub/pdf/scpwps/ecbwp362.pdf?ca7860acb9e0e0a88c32399ddceca956

Exxon Mobil (2016) The outlook for energy: a view to 2040. http://corporate.exxonmobil.com/en/energy/energy-outlook

Financial Times (2013) http://ftalphaville.ft.com/2013/04/24/1469422/the-decline-of-the-oil-spot-market/

Furfari S (2007) Le Monde et l'Energie, Enjeux géopolitiques. Editions Technip, Paris

Greenpeace (2015) Energy revolution. http://www.greenpeace.org/international/Global/international/publications/climate/2015/Energy-Revolution-2015-Full.pdf

IAEA (2014) http://www.oecd-nea.org/ndd/pubs/2014/7209-uranium-2014.pdf

© OECD/IEA (2014) IEA Publishing. License: www.iea.org/t&c. As modified by V. Petit. http://www.iea.org/

© OECD/IEA (2015) Oil market report. IEA Publishing. License: www.iea.org/t&c. As modified by V. Petit. https://www.iea.org/media/omrreports/fullissues/2015-09-11.pdf

© OECD/IEA, Coal Report (2012) Medium term coal market report. IEA Publishing. License: www.iea.org/t&c. As modified by V. Petit. http://www.iea.org/publications/freepublications/publication/mtcoalmr2012_free.pdf

© OECD/IEA, Electricity (2012) Electricity information. IEA Publishing. License: www.iea.org/t&c. As modified by V. Petit. http://www.iea.org/media/training/presentations/statisticsmarch/electricityinformation.pdf

© OECD/IEA, Heating (2014) Heat information. IEA Publishing. License: www.iea.org/t&c. As modified by V. Petit. http://www.iea.org/topics/heat/

© OECD/IEA, Investment (2014) World energy investment outlook. IEA Publishing. License: www.iea.org/t&c. As modified by V. Petit. http://www.iea.org/publications/freepublications/publication/weio2014.pdf

© OECD/IEA, Networks Infrastructures (2013) Electricity networks. IEA Publishing. License: www.iea.org/t&c. As modified by V. Petit. http://www.iea.org/publications/insights/insightpublications/ElectricityNetworks2013_FINAL.pdf

© OECD/IEA, Power Transition (2012) Securing power during the transition. IEA Publishing. License: www.iea.org/t&c. As modified by V. Petit. http://www.iea.org/publications/insights/insightpublications/securing-power-during-the-transition.html

© OECD/IEA, WEO (2012) World Energy Outlook. IEA Publishing. License: www.iea.org/t&c. As modified by V. Petit. http://www.worldenergyoutlook.org/publications/weo-2012/

IMF (2012) International Monetary Fund. Benes J, Chauvet M, Kmaneik O, Kumhof M, Laxton D, Mursula S, Selody J. http://www.imf.org/external/pubs/ft/wp/2012/wp12109.pdf

Infomine (2014) http://www.infomine.com/ChartsAndData/ChartBuilder.aspx?gf=110571.USD.kg&df=19150101&dt=20150111&dr=MAX

Knoema (2014) http://knoema.com/vhzbeig/oil-statistics-production-costs-breakeven-price

Knoema forecast (2015) http://knoema.com/yxptpab/crude-oil-price-forecast-long-term-2015-to-2025-data-and-charts

Macrotrends (2015) http://www.macrotrends.net/1369/crude-oil-price-history-chart

Makortoff (2015) CNBC. http://www.cnbc.com/2015/07/22/world-bank-raises-oil-forecast-but-dont-get-too-hopeful.html

NEI (2014) http://www.nei.org/Knowledge-Center/Nuclear-Statistics/World-Statistics/World-Nuclear-Generation-and-Capacity

Nuclear Energy Agency (2014) http://www.oecd-nea.org/pub/nuclearenergytoday/6885-nuclear-energy-today.pdf

Sharma G (2015) http://www.forbes.com/sites/gauravsharma/2015/07/24/oil-price-will-fall-further-but-2016-17-futures-look-undervalued/3/

Shell (2016) New lens scenarios. http://www.shell.com/promos/english/_jcr_content.stream/1448477051486/08032d761ef7d81a4d3b1b6df8620c1e9a64e564a9548e1f2db02e575b00b765/scenarios-newdoc-english.pdf?

Statoil (2016) Energy market perspectives. Long term macro and market outlook. http://www.statoil.com/no/NewsAndMedia/News/2016/Downloads/Energy%20Perspectives%202016.pdf

Telegraph (2014) http://www.telegraph.co.uk/finance/oilprices/11283875/Bank-of-America-sees-50-oil-as-Opec-dies.html

The New Climate Economy (2016) http://newclimateeconomy.report/2014/finance/

UN/DESA (2014) http://esa.un.org/unpd/wpp/unpp/panel_population.htm

WEC Nuclear (2014) http://www.worldenergy.org/wp-content/uploads/2012/10/PUB_world_energy_perspective__nuclear_energy_one_year_after_fukushima_2012_WEC.pdf

White House (2014) https://www.whitehouse.gov/sites/default/files/docs/the_cost_of_delaying_action_to_stem_climate_change.pdf

World Economic Forum (2015) https://agenda.weforum.org/2015/07/how-low-oil-prices-affect-shale-oil-producers/

World Energy Council (2013) World energy scenarios: composing energy futures to 2050. https://www.worldenergy.org/wp-content/uploads/2013/09/World-Energy-Scenarios_Composing-energy-futures-to-2050_Full-report.pdf

World Nuclear (2014) http://www.world-nuclear.org/info/Nuclear-Fuel-Cycle/Uranium-Resources/Supply-of-Uranium/

WTRG (2013) http://www.wtrg.com/prices.htm

The New Energy Paradigm and Balance of Power

4

Deep historical continuities happen worldwide. The spectacular growth in world population and the economic catching up of new economies lead to a significant increase in energy demand worldwide. Almost everywhere, the economy runs at full speed and the world of energy follows up, using the ways of the past. These continuities are historical determinisms. Economically, they modify deeply the balance of power between countries. Energy-wise, they modify considerably the landscape of geopolitical relationships. The combination of these evolutions either smoothens or accentuates the tensions between nations, in the process redrawing the map of the world's economic centers of gravity. Three major lines of geopolitical tensions are highlighted here: the relative isolation of the American continent, the growing relationships between countries surrounding the Arabian Sea, and the enhanced partnerships between China and the Eurasian shelf.

4.1   North America Isolation

4.1.1   World's Top Energy Consumer

North America consumes 22% of the world's energy even though it is home to just 7% of the world's population. Energy consumption there is 5.9 toe/year/individual, three times more than the world average.

The high energy intensity mainly comes from buildings and transport modes. Consumption in buildings is 1.2 toe/year/individual, three times the world average. The consumption in transportation is 1.5 toe/year/individual, five times the world average. Energy consumption in the transportation sector in North America represents 35% of the total worldwide energy consumption in that sector. This makes North America, and the United States in particular, the top energy consumer in the world.

The consumption should progressively be less intense in the next 20 years. While energy consumption in buildings should continue to grow in absolute

© Springer International Publishing AG 2017
V. Petit, *The Energy Transition*, DOI 10.1007/978-3-319-50292-2_4

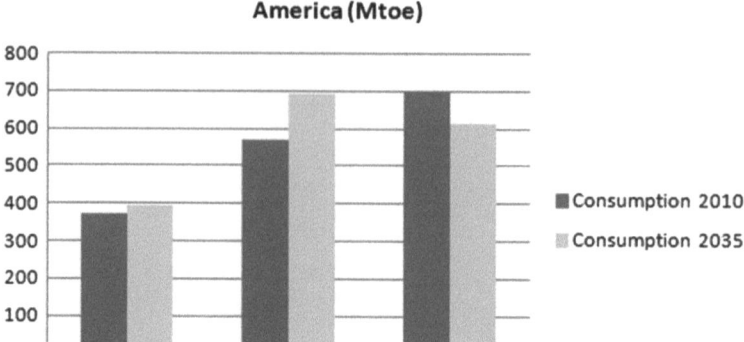

Fig. 4.1 Final energy consumption in North America (© OECD/IEA, WEO 2012)

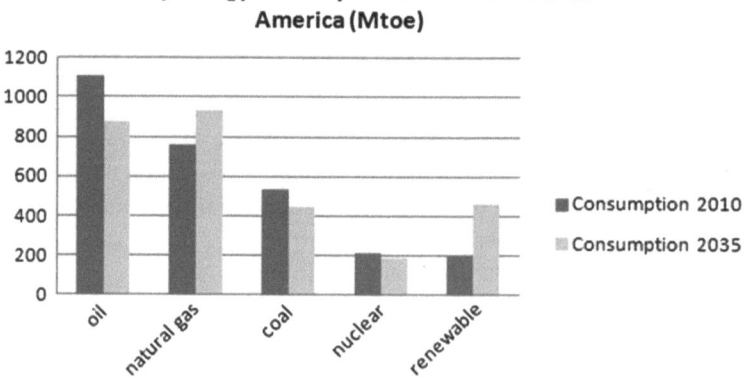

Fig. 4.2 Primary energy consumption in North America (© OECD/IEA, WEO 2012)

value, it should decrease in absolute value in transportation as a result of energy efficiency measures and investments. Energy consumption in North America should drop to 5.1 toe/year/individual by 2035 (Fig. 4.1).

Most of the energy consumed in North America comes from fossil fuels; half of it comes from oil. The dependency on oil should progressively reduce in the coming years as the country initiates energy efficiency measures. The coal footprint would also reduce, while natural gas would increase and take a stronger market share of the overall energy mix of North America. This growth is naturally sustained with local shale gas (Fig. 4.2).

To sum up, North America is by far the topmost energy consumer in the world, globally and per individual. This shall continue to be the case, even though the gap with other regions will narrow. In absolute value terms, the energy needs of the region shall continue to increase because of the dynamics of its population and of its economy, but the growth would be tamed thanks to energy efficiency measures, in particular in transportation.

4.1.2 The Energy-Independent Continent

North America consumes more than 2800 billion tons of oil equivalent per year. It produces 75% of its energy needs and is thus almost energy-independent. The remaining 25% is imported (mostly as oil) from South America and the Middle East. Natural gas imports are negligible. North American thus has no critical dependency on any other region of the world (Fig. 4.3).

Whether this independence can be sustained in the long term is a matter of concern.

Figure 4.4 maps the ratio of reserves to production for each source of energy (in years of consumption). R/P uses retrievable reserves and Rprov/P shows the same ratio for proven reserves. The 2035 ratio is calculated using the current production of 2035 and the calculated cumulated depletion of retrievable reserves by 2035. The ratio of production to consumption is also shown to allow assessment of North America's current energy independence per type of energy. The 2035 forecast is the one from the International Energy Agency (2012).

"Proven" reserves of North America amount up to 40 years of oil and 15 years of natural gas. Coal reserves correspond to 300 years of production at the current pace. The discovery of shale oil and gas as well as other unconventional oils in Canada has considerably modified the status of reserves in the region. "Retrievable" resources in North America are now the highest in the world, with more than 400 years of oil production at the current pace and 140 years of natural gas. Only 10% of these reserves are "proven" but North America is theoretically independent from an energy standpoint. The development of these resources, provided they are economically sound, would lead to a slight reduction of the R/P ratio, but the energy

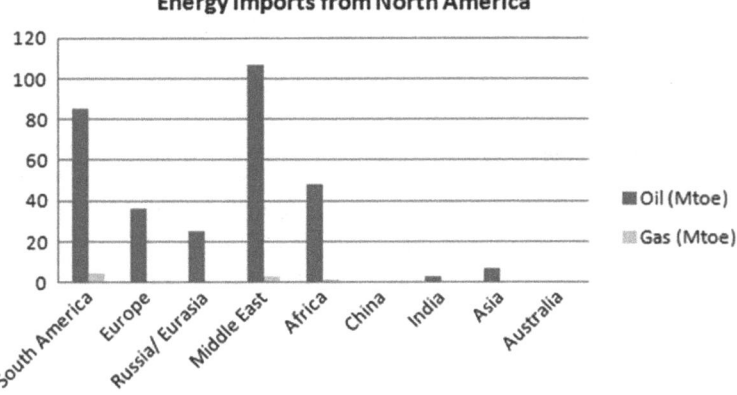

Fig. 4.3 Energy imports from North America (BP 2009, 2014; © OECD/IEA, WEO 2012)

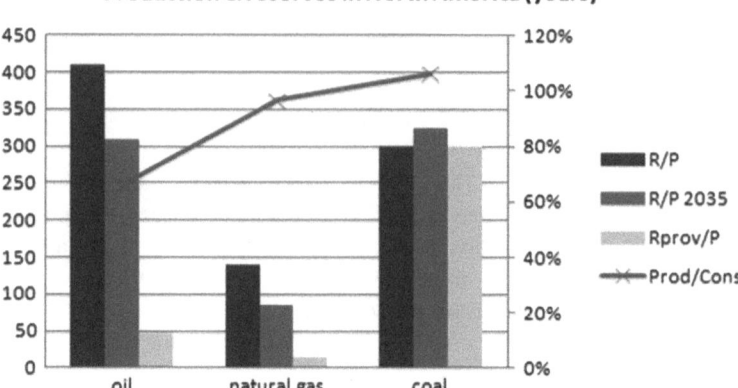

Fig. 4.4 Production and Reserves in North America (BP 2014; © OECD/IEA, WEO 2012)

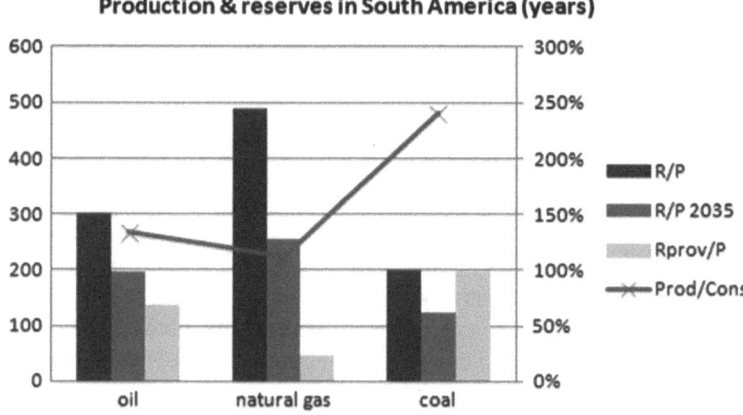

Fig. 4.5 Production and Reserves in South America (BP 2014; © OECD/IEA, WEO 2012)

independence of the region would not be put at risk in the short term, both for oil and natural gas.

South America also has important fossil fuel reserves. "Proven" reserves represent 140 years of oil production and almost 50 years of natural gas at the current pace. The continent has about 300 years of "retrievable" oil reserves and 500 years of natural gas. South America is thus completely autonomous from an energy standpoint and an attractive complement of resources to North America. Despite a projected 50% growth in energy consumption in the next 20 years, the reserves will be depleted at a low rate with regards the total volume available. Still, there is a need to convert retrievable reserves into "proven" reserves. A high enough oil price (or natural gas price) will be required to trigger this (Fig. 4.5).

The American continent as a whole (i.e. North America and South America) thus represents 35% of global fossil fuel resources, 25% of natural gas and 50% of oil reserves, while its population corresponds to only 14% of the world population. Twenty years back, its identified resources were twice lower.

This evolution of scale is clearly a change of paradigm for the continent and should impact the way it (particularly North America) deals with other regions of the world.

Given that North America theoretically has over 400 years of oil production reserves and that South America also shows spectacular potential, the United States energy policy could evolve radically. Of course, the historical presence of the United States in the Middle East would not change dramatically in the coming few years. As well, "proven" reserves only amount up to 10% of the total of "retrievable" reserves. If a sustained price war was to last between the United States (shale oil producers) and the Middle East (Saudi Arabia in particular), the foreign policy of the United States in the region could be modified. With 10% of total energy consumed in North America, the market share of the Middle East has significantly dropped in the past few years, together with its influence. The consequences of this fact of oil geopolitics remain to be written.

4.2 The Enlarged Arabian Sea

4.2.1 Oil from the Arabic Peninsula

4.2.1.1 An Economy Strongly Dependent on Oil

Energy production in the Middle East represents 16% of global production. In the area of oil production alone, it accounts for 32% of global output; the region produces four times more oil than it consumes. Not surprisingly, the Middle East is the heart of world oil exchanges.

The well explored region has 80 years of "proven" reserves at the current pace of production and 170 years of natural gas. Its "proven" reserves represent 60% of its total "retrievable" resources. With them, the region has 120 years of oil production and close to 300 years of natural gas production. The region thus has sustainable access to resources over the long term and can continue to produce at the current pace and possibly increase its output in the coming decades. Actually, the world has fueled its growth with oil from the Middle East for decades, a region often qualified as "a geological anomaly" (Chevalier 2012). According to forecasts from the International Energy Agency (2012), the region would still have in 2035 over 50 years of reserves of oil and 150 years of reserves of natural gas. However, its reserves of coal are extremely low and would not allow it to supply coal to China for more than 6 months; coal production is virtually non-existent in the region (Fig. 4.6).

The share of oil exports from the Middle East should continue to increase in the coming years, to reach around 36% of global oil production (versus 32% today). This is related to the low cost of production of oil in the region, the most competitive in the world. Actually, the Middle East region is always favored thanks to its oil

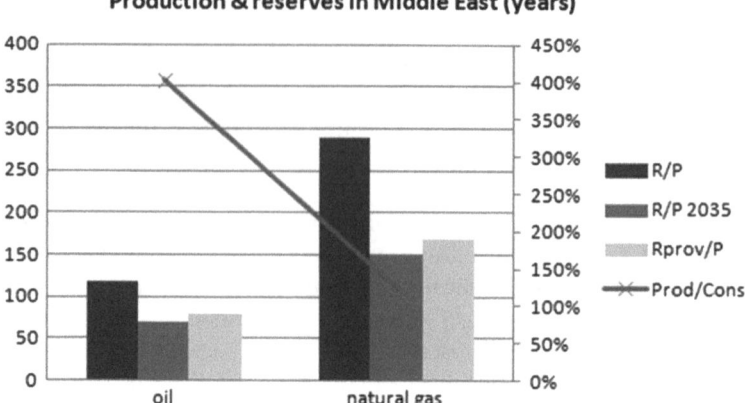

Fig. 4.6 Production and Reserves in the Middle East (BP 2014; © OECD/IEA, WEO 2012)

production cost. If the oil price is high, investments in other regions of the world become profitable, but the Middle East countries benefit from an important oil rent. When the oil price drops, the rent decreases, but investments in other regions can be frozen because they are not profitable anymore. The energy exports of the Middle East vary strongly from one region to another. Asia (including China and India) today represents 75% of its exports. The "influence ratio" compares the total volume of oil exports from the Middle East in a given region to the overall total consumption of oil. It indicates the actual influence the Middle East can exert on a given region. It shows that 70% of oil consumption in India and 60% in Asia (excluding China) depend on Middle East supplies. The relations between these three regions will thus remain intricate in the coming years. In particular, the oil consumption boom in India as well as its historical relationship with the Gulf countries could lead the country to reinforce its political and economic ties with the Middle East in the coming years (Fig. 4.7).

Beyond this specific situation, the economy of the Middle East depends highly on the revenues of its oil industry. These revenues represent indeed 75% of the government's budgets in Saudi Arabia, Qatar and the United Arab Emirates. In Iran, they represent more than 50% of the government budget. The economy of the region is highly dependent on oil price.

4.2.1.2 The Arabic Peninsula Economy

Saudi Arabia was the first economic power of the region, with 750 billion dollars of GDP in 2013 (World Bank 2014), four times that in 2000. More than half (55%) of its GDP comes directly from the oil industry. The remaining 45% are related to the private sector, which consists mainly of the construction sector, which is highly dependent on government spending and therefore on oil price.

Saudi demographics are very dynamic, even though the birth rate dropped in the last 30 years to reach around three children per family today. The country is

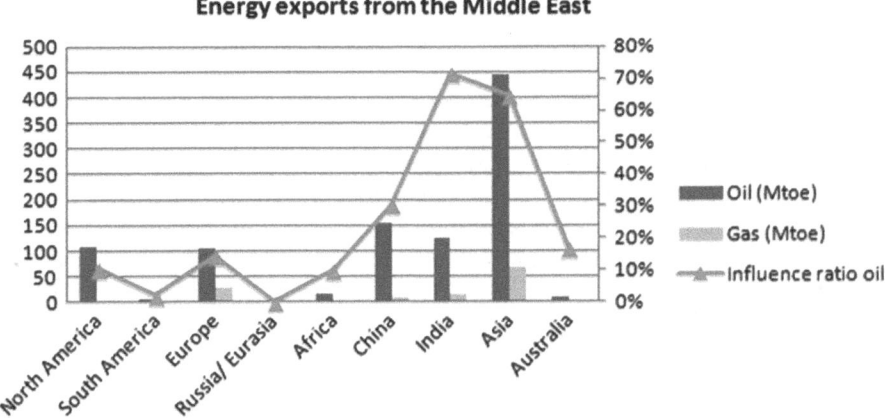

Fig. 4.7 Energy exports from the Middle East (BP 2009, 2014; © OECD/IEA, WEO 2012)

therefore in the middle of its demographic transition, and the drop in infant mortality rate, as well as the increase in life expectancy, led to a sharp increase of the population. More than 50% of the population of the country is less than 25 years old (Murphy 2012).

By 2018, while more than 1.6 million locals will enter the job market in the region, only half would be able to find a job (Associated Press 2013). Almost 90% of the seven million jobs created in the region between 2000 and 2010 were taken up by foreigners. In Saudi Arabia, one third of the 25 million inhabitants is a foreigner. The rather weak development of education is the main root cause of this situation. A few years back, the government tried to impose on private companies that they employ at least 30% of Saudi nationals; an outcry from the private industry caused it to backtrack. Currently, while there have been investments from the government in education, they have not yet paid off. As a result, around 30% of the population aged 20–30 years would today be unemployed (The Economist 2012).

According to The Guardian (2013), this situation puts 25% of the population below the poverty line. Only 30% of Saudis own their own house; the world average is 70% (Sloan 2014).

As a result, the Saudi economy is highly dependent on oil rent. The government needs an oil price of around 100 dollars a barrel to balance its budget; its incentive cost of production is much lower, at around 27 dollars per barrel (Energy Matters 2014; Knoema 2014). Today, 79% of the active population is tied economically to government activities.

Saudi Arabia has around 700 billion dollars of currency reserves and can therefore face difficult times. Still, a sustained oil price below the budget breakeven level could lead to unexpected evolutions.

To a lesser extent, the United Arab Emirates, Qatar and other countries in the peninsula are in the same situation. Their budget strongly depends on oil revenues, their population is increasing rapidly, the employment rate of young people remains low, and a significant share of their economy relies on foreign workers.

4.2.1.3 Oil Price and Political Stability

The spectacular development of unconventional oil and gas in North America, as well as the recent discoveries in South America, have resulted in a world where energy resources are abundant. This presents a paradox. Oil price tends to drop when there is no shortage of capacity, turning investments into non-profitable ones. In addition, the Asian economies which drive energy demand and maintained for years a high level of oil price are slowing down. In the end, the paradigm for the last 20 years has brutally changed. Growth is weaker, energy consumption is growing at a slower rate, and reserves are abundant.

Now, this massive amount of retrievable reserves can be developed into "proven" reserves provided the price reaches a sufficiently high incentive level for investors. The Middle East region (and OPEC) is thus likely to stay a dominant player on the market, although its actual ability to control the price level and the market is undermined by the massive input of unconventional oils as well as its significantly reduced influence over large consumers such as North America, Europe and China. The "Quincy" agreement (named after the warship on which Roosevelt signed the agreement in 1945 with Ibn Saud, the first king of Saudi Arabia) which settled the long-lasting relationship between the United States and the kingdom of Saudi Arabia, renewed by George W. Bush in 2006, could have endured. It appears that Asian countries are now much more tied to the oil from the Middle East than North America or Europe. The consequences of this spectacular turnaround are yet to be seen on the political scene. In the meantime, a sustained oil price below the budget breakeven level of the Middle East region could lead to increased tensions in the Middle East countries as it would hamper the ability of these countries to build a more diversified economy. The fact that the bulk of unconventional oil from North America has incentive costs below 80 USD/bl suggests that, in the mid term, prices are likely to stay well below 100 USD/bl, the price at which most Middle East governments balance their national budgets.

In the end, the Middle East region still owns almost half of the world's "proven" reserves of oil, which many countries are highly dependent on. As the Indian or other Asian economies reach new levels, their political influence is likely to be stronger abroad. The Middle East will certainly be one of their first destinations.

4.2.2 India, the Great Partner

4.2.2.1 Indian Economy Takes Off

India's economic take-off was long awaited. Gurcharan Das (2000) considers Nehru's policy of the "iron frame" in the 1950s as the primary cause of the delay. Although the situation has considerably improved in recent years, a number of issues still persist today. Infrastructures remain underdeveloped, daily electricity supply interruptions are common in large cities, water networks are often deficient, and health services vary significantly from one place to another. According to the World Bank (2014), 36% of the population benefit from access to improved sanitary conditions. There are fewer than 0.7 doctors for every 1000 inhabitants, malnutrition remains a curse for millions of people in the country, more than 17% of the population is undernourished, and 43% of children younger than 5 years old are malnourished. Despite these worrying figures, the situation is better than 40 years ago, when 25% of the population (and 70% of the children) were undernourished.

Conditions have improved significantly since the reforms of 1991. India's GDP grew 5% on average the last 10 years (Maps of India 2014). Population growth is about 1.2% per year, after reaching a peak 2.5% in 1970. The birth rate has dropped from six children per woman in the 1960s to 2.5 today. Life expectancy rose from 40 years in 1960 to 66 years today.

This spectacular economic growth is based on very specific principles. Unlike China, which focused on manufacturing products for export, India chose to focus on domestic consumption and the development of services. More than half of its GDP is today generated by services. The industry sector represents only 27% of total GDP. This policy led India to maintain a Gini coefficient at a reasonable level, around 33, comparable to that of Germany (30). It is lower than that of China (39), the United States (41) and Brazil (59) (Gapminder 2014). Actually, the weak development of the industry sector in India is also the primary result of a remarkable culture of entrepreneurship. An impressive 82% of people in India are self-employed (World Bank 2014), against 7% in the United States and 11% in France. The country's economy is therefore based on small-business local entrepreneurship. This particularity dates back ages, and was reinforced by the different measures taken by governments which ruled India in the past decades, in particular the ones headed by Indira Gandhi (between 1966 and 1984), which prevented large companies from taking over control of the manufacture of products. This led to the development of an economy centered around its own internal market and on the development of services. This made of India the world's "back office" (Das 2000).

The Indian economy took off recently. With 5% GDP growth annually, a high birth rate, a large share of young people in its population, increasing life expectancy, a developing industry sector, modernizing bureaucracy and improving infrastructures, India will become in the twenty-first century a great champion of growth and a major power in the world balance.

This giant requires energy resources to fuel its growth.

4.2.2.2 Growing Energy Needs

India is home 17% of the world population but only accounts for 3% of global GDP and 6% of the energy consumption worldwide. The consumption per individual is one of the lowest in the world at 0.5 toe/year/individual.

Energy consumption per individual in the buildings sector is the lowest in the world at only 0.16 toe/year/individual. This is linked to the level of urbanization, which barely tops 30%. Energy consumption is essentially based on biomass and waste (71% of the total energy consumed). Energy consumption by transportation is also the lowest in the world, almost seven times less than the world average. Three

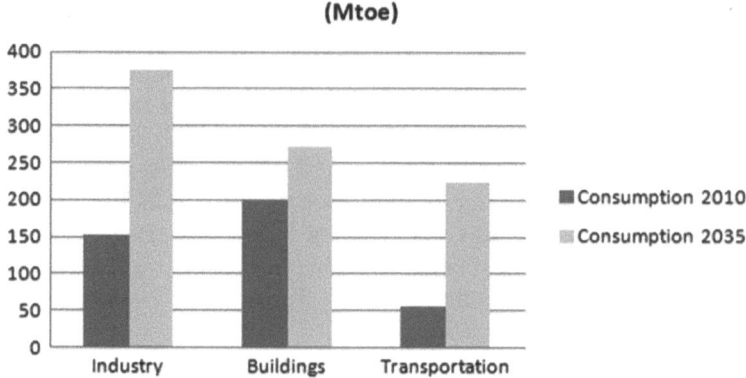

Fig. 4.8 Final energy consumption in India (© OECD/IEA, WEO 2012)

quarters of all travel are done by bus or rail, against almost 20% in Europe and only 4% in North America, the most heavily oil-dependent regions.

Within 20 years, the energy consumption of India should more than double and represent 9% of the energy consumption worldwide (against 6% today). Individual consumption should also double to about 1 toe/year/individual, which will still remain half of the world average of 1.9 toe/year/individual in 2035 (against 1.7 toe/year/individual today). A large portion of the increase in the country's energy consumption will come from the development of its industry (48% of the total increase) as well as transportation (36% of the total increase) (Fig. 4.8).

Primary energy consumption remains overall very dependent on coal, which corresponds to 41% of the total. It is one of the highest ratios in the world, although lagging that for China. India also produces the majority of the coal it consumes. Renewable energy is also an important part of the energy mix of the country, and is widely used for heating and cooking. Within the next 20 years, energy consumption in India will more than double. Oil and coal consumption shall double, while natural gas consumption shall be multiplied by three. The share of nuclear power shall double in the mix, which means nuclear-based electricity production would be multiplied by six in 20 years (© OECD/IEA 2014) (Fig. 4.9).

4.2.2.3 The Energy Challenge of India

India consumes much more energy than it produces. It has important coal reserves equal to 175 years of production at the current pace of consumption. In absolute value, Indian coal reserves are twice lower than Chinese reserves. In the short term, India will sustain its development with its coal reserves. However, as production increases to meet consumption, it should only have 50 years of production left by 2035. The country would then face a new energy challenge.

Besides coal, India does not have reserves of oil or natural gas. This will limit its development. The country is essentially dependent on oil, with around 80% of its oil consumption imported, mostly from the Middle East. This is one of the highest

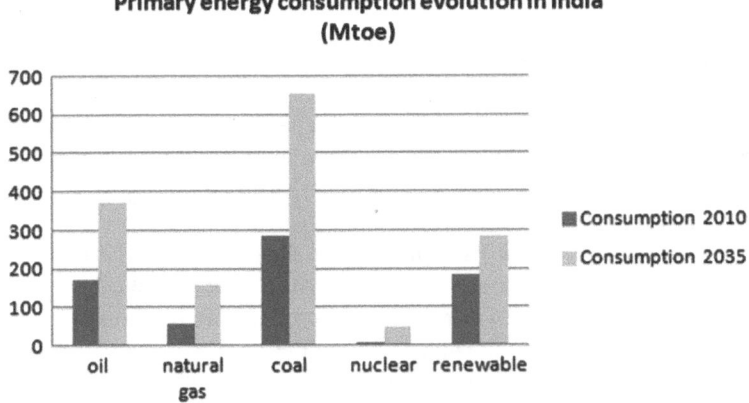

Fig. 4.9 Primary energy consumption in India (© OECD/IEA, WEO 2012)

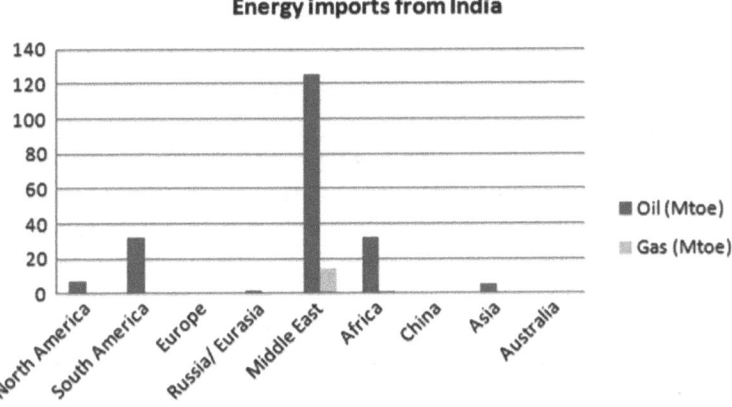

Fig. 4.10 Energy imports from India (BP 2009, 2014; © OECD/IEA, WEO 2012)

dependency ratios of the world. India thus highly depends on its neighbors and on the rest of the world to keep up with growth. Its long-term growth will only be sustained if it maintains long-term stable partnerships with a number of regional powers, in particular the Middle East. As India will likely take a higher share of the oil exports from the region in the coming years, its political and economic ties are expected to progressively become stronger (Fig. 4.10).

4.3   The Border Between China and Russia

4.3.1   China Gets Back on Top

China consumes almost a fifth of the total energy consumed in the world. It contributes 17% of global GDP and is home to around 19% of the world population. The individual consumption averages 1.8 toe/year/individual, just slightly above the world average.

The country is first an industrial giant. More than half of its energy needs come from industry, by far the highest ratio in the world, the average being 33%. For 20 years, China has been the world's "workshop". Its buildings and transportation sectors' energy consumption is slightly below the world average, around 0.3 toe/year/individual for buildings (compared to 0.4 toe/year/individual world average) and 0.14 toe/year/individual for transportation (compared to 0.23 toe/year/individual world average). The main root cause of China's energy intensity is thus related to industry, in both volume and percentage terms.

Energy consumption in China should increase by more than 50% within the next 20 years to reach 22% of total energy consumption worldwide. This growth will be related to both the continual development of its industry as well as the rising living standards of its population. By 2035 China shall be the topmost energy consumer in the world, far ahead of North America (© OECD/IEA 2014; © OECD/IEA, WEO 2012). Energy consumption per individual will rise to around 2.6 toe/year/individual, slightly below Europe, but much more than the world average, which should equal 1.9 toe/year/individual in 2035 (Fig. 4.11).

Most of China's primary energy consumption is based on fossil fuels. The country alone consumes half of the world's coal production; renewable energies and nuclear energy only represent 13% of total energy consumption. The Chinese energy mix should evolve in the coming 20 years. Its total energy consumption shall increase by more than 50%. Oil consumption would increase by 70%, while coal

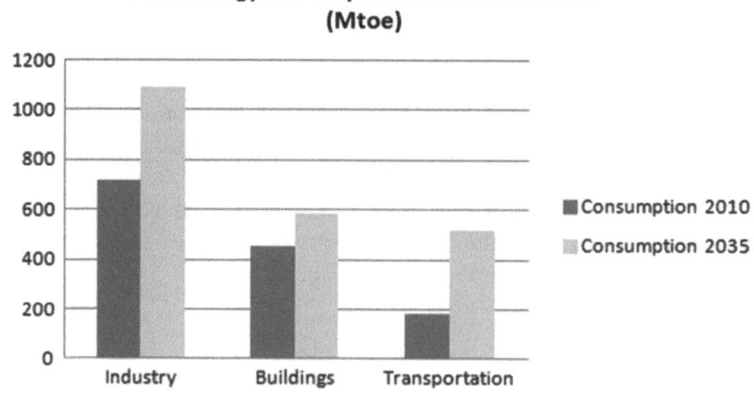

Fig. 4.11 Final energy consumption in China (© OECD/IEA, WEO 2012)

consumption shall only increase by 23%. Nuclear production would be multiplied by six, natural gas consumption by five and total renewable energy production shall be two times higher than today. The share of coal in the total energy mix of China would then drop below 50% of the total energy consumption of the country (Fig. 4.12).

Actually, the main issue of China is the status of its reserves. The country has very few oil resources. Its "proven" reserves of natural gas are also small, but it would have the first "retrievable" unconventional gas reserves in the world (EIA 2013). It imports massively from other regions of the world the oil and natural gas it requires. In particular, it imports 30% of the oil it consumes from the Middle East. Although this percentage is much lower than that for India or the rest of Asia, China is very dependent on oil from the Middle East. Geographically, the country is also very close to the Eurasia region (Russia, Kazakhstan, Uzbekistan, etc.). The level of oil (and natural gas) imported today is however very small in absolute value as it does not exceed 8% of total oil consumption (and 17% of total natural gas consumption) (Fig. 4.13).

As for coal, China is the world's biggest producer, with almost 50% of global production. At the current pace of production, China has 50 years of production left. Considering the steep growth of its consumption (which doubled in the last 10 years, and should increase by another 23% by 2035), the current depletion of reserves should lead to less than 19 years of production left by 2035. China thus has to make major changes to its energy mix or lose significant energy autonomy.

Eventually, China will have to explore the use of unconventional gas and renewable energies as components of a wider energy mix. Natural gas and oil from Siberia also offer potential; these resources are today underdeveloped in light of the massive needs of China (Fig. 4.14).

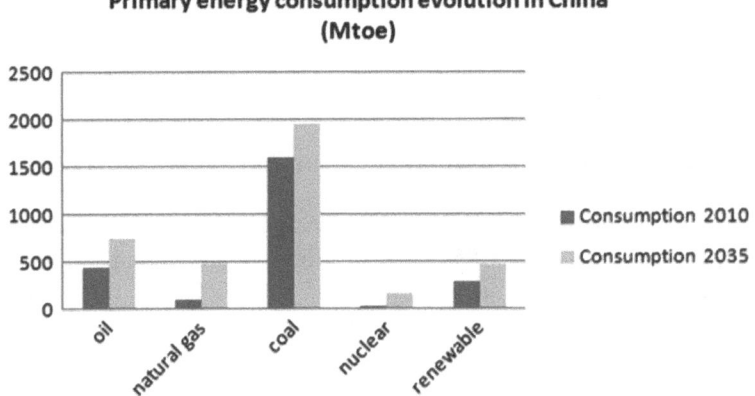

Fig. 4.12 Primary energy consumption in China (© OECD/IEA, WEO 2012)

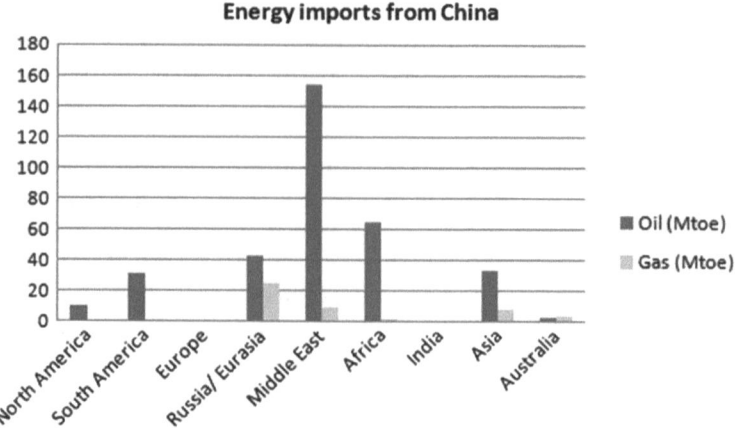

Fig. 4.13 Energy imports from China (BP 2009, 2014; © OECD/IEA, WEO 2012)

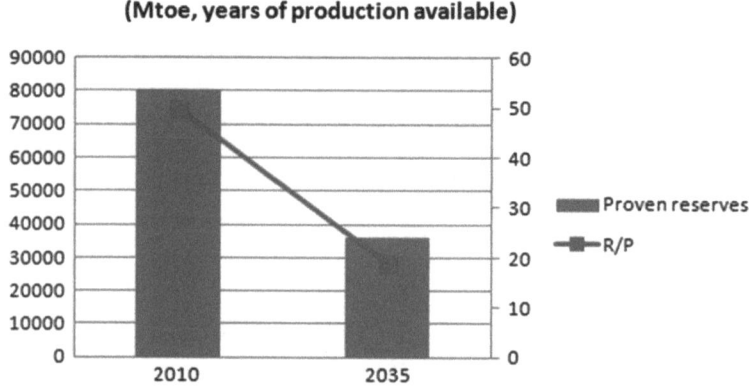

Fig. 4.14 Coal reserves' depletion in China (BP 2014; © OECD/IEA, WEO 2012)

4.3.2   Strengths and Weaknesses of the Eurasian Continent

Eurasia is a very large producer of primary energy resources. The 2000 Mtoe it produces per year gives it an 18% share of global production. The region is the third-largest oil producer and the biggest natural gas producer in the world (on par with North America). It produces twice more energy than it consumes, and has up to 25 years of oil and 60 years of natural gas in proven reserves at the current pace of production. Provided the massive investments required to put in operation its retrievable resources are triggered, Eurasia would have around 200 years of production for both oil and natural gas. The investments would add up to 3.2 trillion dollars within the next 20 years (© OECD/IEA, Investment 2014). In comparison, North America plans to invest around 6.4 trillion dollars in primary energies by

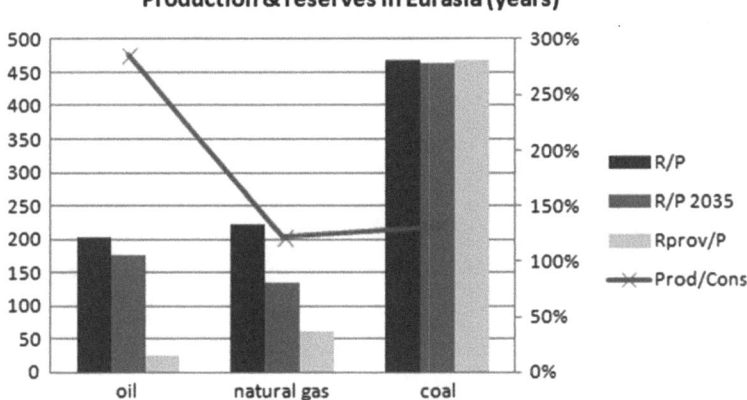

Fig. 4.15 Production and Reserves in Eurasia (BP 2014; © OECD/IEA, WEO 2012)

2035. Eurasia also has 470 years of coal production at the current pace of production (equivalent to a little more than 10 years of production in China), and 60 years of uranium, although these reserves are difficult to evaluate because exploration has so far remained limited. Kazakhstan is the leading producer of uranium in the world.

The region has therefore very large reserves of fossil fuel resources. All energies included, it has more resources than the Middle East, and comes second only after North America. Taking into account the growth in its production in the coming years (© OECD/IEA, WEO 2012), its ratio of reserves to production (R/P) should not evolve significantly, with the exception of natural gas, which should drop below 150 years of production, a level which remains comfortable (Fig. 4.15).

Eurasia is a net exporter of energy to the rest of the world. In particular, it exports both oil and natural gas to Europe, with which it has strong economic ties. The "influence ratio" does not exceed 40%, which means that the exports from Eurasia do not exceed 40% of Europe's total consumption. Europe has been working at reducing this ratio in the past years and is diversifying its procurements. There are as well major differences across countries in terms of energy dependency, with Eastern European countries traditionally much more dependent on Eurasia than countries in the west of Europe. The influence that Eurasia can have over Europe in energy remains therefore limited (Fig. 4.16).

In summary, the Eurasia region is traditionally very connected to Europe, and much less to China, which growing needs are spectacular. Provided the level of investments required can be justified, it is expected that the relationship between China and Eurasia countries (particularly Russia) will be reinforced in the coming years. This would help China diversify its energy procurements and enable its energy transition out of coal. Natural gas from the eastern regions of Russia could play a significant role in this transition.

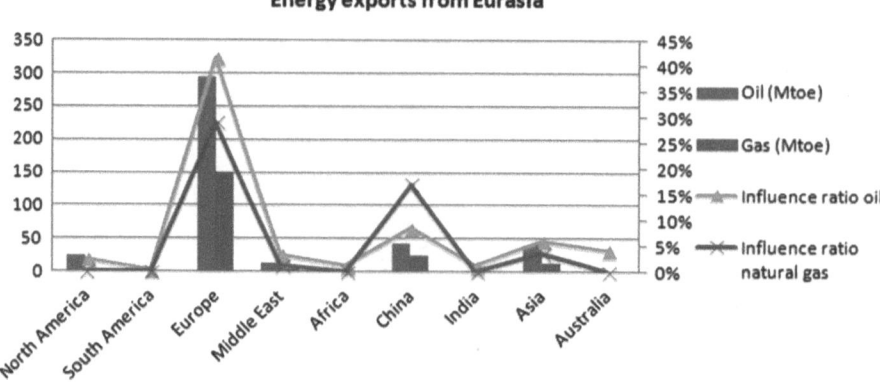

Fig. 4.16   Energy exports from Eurasia (BP 2009, 2014; © OECD/IEA, WEO 2012)

References

Associated Press (2013) http://www.foxnews.com/world/2013/12/03/saudi-arabia-needs-private-sector-growth-to-stem-youth-unemployment-business/

BP (2009) http://www.jaimelafinance.com/2011/02/petrole-quels-sont-les-flux-mondiaux.html

BP (2014) BP statistical review (on website)

Chevalier JM (2012) Les grandes batailles de l'énergie. Folio, Paris

Chevalier JM, Derdevet M, Geoffron P (2012) L'avenir énergétique: cartes sur table. Folio, Paris

Das G (2000) Le réveil de l'Inde. Editions Buchet Chastel, Paris

EIA (2013) Technically recoverable shale oil and shale gas resources: an assessment of 137 shale formations in 41 countries outside the united states. Energy Information Agency. http://www.eia.gov/analysis/studies/worldshalegas/

Energy Matters (2014) http://euanmearns.com/oil-price-wars-who-blinks-first/

Gapminder (2014) http://www.gapminder.org/world/#$majorMode=chart$is;shi=t;ly=2003;lb=f;il=t;fs=11;al=30;stl=t;st=t;nsl=t;se=t$wst;tts=C$ts;sp=5.59290322580644;ti=1899$zpv;v=0$inc_x;mmid=XCOORDS;iid=phAwcNAVuyj1jiMAkmq1iMg;by=ind$inc_y;mmid=YCOORDS;iid=phAwcNAVuyj2tPLxKvvnNPA;by=ind$inc_s;uniValue=8.21;iid=phAwcNAVuyj0XOoBL_n5tAQ;by=ind$inc_c;uniValue=255;gid=CATID0;by=grp$map_x;scale=log;dataMin=283;dataMax=110808$map_y;scale=lin;dataMin=18;dataMax=87$map_s;sma=49;smi=2.65$cd;bd=0$inds=;modified=75

© OECD/IEA, WEO (2012) World Energy Outlook. IEA Publishing. License: www.iea.org/t&c. As modified by V. Petit. http://www.worldenergyoutlook.org/publications/weo-2012/

© OECD/IEA (2014) IEA Publishing. License: www.iea.org/t&c. As modified by V. Petit. http://www.iea.org/

© OECD/IEA, Investment (2014) World energy investment outlook. IEA Publishing. License: www.iea.org/t&c. As modified by V. Petit. http://www.iea.org/publications/freepublications/publication/weio2014.pdf

Knoema (2014) http://knoema.com/vhzbeig/oil-statistics-production-costs-breakeven-price

Maps of India (2014) http://www.mapsofindia.com/economy/

Murphy C (2012) http://www.newsecuritybeat.org/2012/02/saudi-arabias-youth-and-the-kingdoms-future/

Sloan A (2014) https://www.middleeastmonitor.com/articles/middle-east/10030-saudi-arabias-housing-predicament

The Economist (2012) http://www.economist.com/node/21548973

The Guardian (2013) http://www.theguardian.com/world/2013/jan/01/saudi-arabia-riyadh-pov
  erty-inequality

World Bank (2014) http://databank.worldbank.org/data/home.aspx; http://data.worldbank.org/
  indicator/SP.URB.TOTL.IN.ZS

The Path Towards a Sustainable Tomorrow

5

As its population increases and living standards improve, the world reaches new heights in terms of energy needs. The energy industry works at full speed to meet a very inelastic demand. Everywhere, it is pushing to the limits the technology that it developed decades ago, while looking with excitement to new ways of doing the same thing. The basics of the processes which drove the Industrial Revolution have been perfected but their principles have so far remained unchallenged. Consequently, the world continues to spend tremendous amounts of energy with massive inefficiency. The acceleration of population growth and the accompanying higher need for energy have surfaced new issues. More and more people are beginning to wonder if the world is able to meet its growing needs, and for how long more? Progressively, people are realizing that they are creating an energy debt for future generations and that this debt will be what they will remember us for.

5.1 The Impossible Energy Equation

A spectacular transformation happened in the last 70 years that was not fully perceived by the general public. On one hand, the world population tremendously increased. The world population, which stood at 2.5 billion in 1950, now tops 7.6 billion, and should reach nine to ten billion by 2050. It would have thus been multiplied by four within one century. On the other hand, propelled by the benefits from successive industrial revolutions, the world population has been benefiting in the last century from a spectacular increase in living conditions, thanks to access to modern health services, food, clean water, and energy. Industrial revolutions have also brought wealth and opportunity to populations that were previously isolated. This has led to a massive increase in energy consumption. Overall, primary energy demand grew more than 45% in the last 20 years. This increase is just the beginning.

Global energy consumption is expected to continue to increase by almost 35% by 2035 (and 50% by 2050). The end-use energy consumption by industry is set to

© Springer International Publishing AG 2017
V. Petit, *The Energy Transition*, DOI 10.1007/978-3-319-50292-2_5

grow 45% within this period, in buildings by 30%, and in transportation by almost 40%. Fossil fuels feed this spectacular increase in demand. On average, their consumption is set to increase by 25% by 2035. Oil shall increase by 15%, coal by 20% and natural gas by almost 50%. Electricity consumption, based on fossil fuels for 75% of its mix, will increase by more than 70% within this time period.

When burned, fossil fuels emit a variety of greenhouse gases which accumulate in the atmosphere. CO_2 emissions take up the main share of greenhouse gas emissions. These emissions can stay within the atmosphere for as long as 200 years. Therefore, the concentration builds up as emissions continue to be released. The concentration of CO_2 has risen by more than 40% in a century (OMM 2014). It is now 400 ppm, far higher than 280 ppm, the highest reading during the last 600,000 years (Durand 2007). According to the GIEC/IPCC (2007), emissions have grown by 70% in the last 70 years.

There is unanimous consensus in the scientific community that the increase in CO_2 concentration will lead to unpredictable climatic changes. The detailed conclusions from the IPCC are supported by most national science academies in the world (National Academies 2009) and there is actually no debate within the scientific community on the urgent need to reduce CO_2 concentration in the atmosphere, even though some scientists continue to question the models used by IPCC to relate greenhouse gas emissions to world temperature change.

The 450 scenario aims to limit the CO_2 concentration level to 450 ppm in the coming decades (© OECD/IEA, WEO 2012). This scenario was developed to evaluate what it would require to actually prevent too high a concentration of CO_2 in the atmosphere. According to the GIEC/IPCC (2007), this concentration would limit world temperature increase to two degrees. Achieving this scenario would require a massive reduction of CO_2 emissions. CO_2 emissions top today 32 Gt/year (© OECD/IEA, WEO 2012; © OECD/IEA, ETP 2012) and need to drop to 22 Gt/year by 2035 to achieve this scenario. Current forecasts, however, point to an increase of up to 44 Gt/year. While energy consumption is set to increase by 35% in the coming 20 years, fed by fossil fuels, the 450 scenario requires a drop of CO_2 emissions of 35% within the same period. It is however currently set to mechanically increase by the same percentage over the period. This is the impossible energy equation which the world will face in the next two decades. Solving the equation will require that we deeply modify the way we use energy, and likely also the way humanity considers the world it lives in.

The "New Policy" scenario (© OECD/IEA, WEO 2012) estimates that primary energy demand worldwide will grow 35% from a 2010 base of 13,000 Mtoe. The 450 scenario would be viable if 2400 Mtoe of primary energy demand were saved out of this growth. This corresponds to taming primary energy demand growth by more than 50%. This is the target to reach.

5.2   Massive Potential for End-Use Efficiency

Each end-use sector presents different opportunities for saving energy. Indeed, increased energy consumption in a number of sectors will be inexorable, while in some others, savings will be easier to reach.

Despite a 35% growth forecast for energy demand in the coming 20 years, the energy saving potential across all end-use sectors amounts to around 25% of the total final energy consumption. The International Energy Agency (2012) indeed records the actual savings forecasted to be realized by 2035 by sector. Those savings, according to the same source, are expected to correspond to 42% of the full potential in the industry sector, 18% in buildings, and 38% in transportation. This yields a full energy efficiency potential which represents as of today 25% of the final end user consumption in the industry sector, 20% in buildings, and 31% in the transportation sector. Overall, this potential corresponds to 1900 Mtoe of end-use energy, meaning that up to 2700 Mtoe of corresponding primary energy could be saved (© OECD/IEA 2014; © OECD/IEA, WEO 2012). Indeed, primary energy can be retrieved using coefficients of transformation for electricity production (40% average yield) and for vehicle gas transformation in refineries (83% of yield). 2700 Mtoe correspond to 21% savings on overall primary energy consumption. This figure needs to be compared to the 450 scenario which plans for a 2400 Mtoe reduction of primary energy consumption by 2035. The 450 scenario thus corresponds to an 88% realization of the total potential in terms of end-use energy efficiency (© OECD/IEA, Energy Efficiency 2013; © OECD/IEA, WEO 2012) (Fig. 5.1).

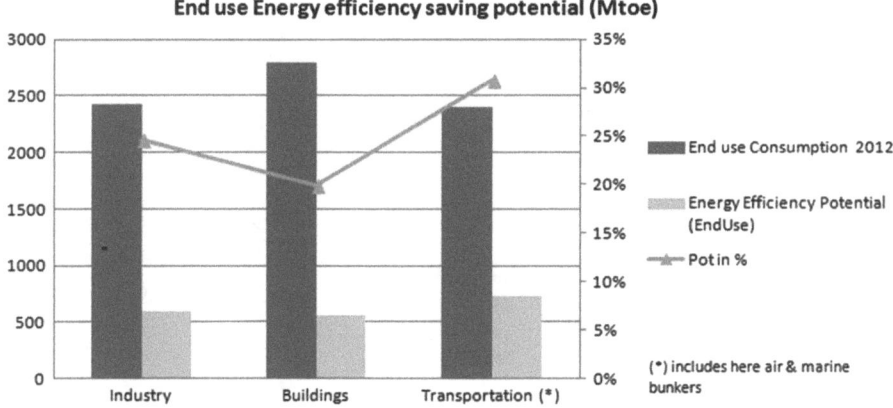

Fig. 5.1 End use energy efficiency saving potential (© OECD/IEA, Energy Efficiency 2013; © OECD/IEA, WEO 2012)

5.2.1   Energy Efficiency Potential in Industry Sector

The bulk of energy-intensive industries include the iron, steel, metals, minerals and petrochemical industries. Together, they account for 50% of the total energy consumed by industry. Figure 5.2 below shows the variation of energy consumption between 2012 and 2035 (© OECD/IEA, Statistics 2015; © OECD/IEA, WEO 2012).

Reducing energy consumption in the industry sector can obviously be achieved by increasing the efficiency of all industries. There, the basic application of the most efficient technologies would lead to massive savings. The challenge of energy efficiency in the sector mainly depends on existing solutions. It can also be achieved by limiting (and then reducing) the energy consumption in energy-intensive industries, in particular, petrochemical industries.

5.2.1.1 Petrochemical Plants

Experts agree that up to 30% of the energy consumed could be saved in the petrochemical industry by using high-performing technologies (© OECD/IEA, Technology Industry 2009). These savings can be grouped into three main technology domains. First, optimizing the heat produced through process integration could lead to savings of around 5–10%. Then, using Combined Heating and Power plants (CHP) to supply petrochemical plants would save up to 20% on existing sites and between 5 and 10% on new sites. Combined Heating and Power directly reuses the heat generated by the production of electricity; in conventional supply systems, electricity is partially used to produce heat. Finally, recycling waste plastics (which are based on oil) could lead to 3% resource savings.

The deployment of these technologies however faces the reality of existing investments and the lifetime of existing sites, which obviously limit immediate

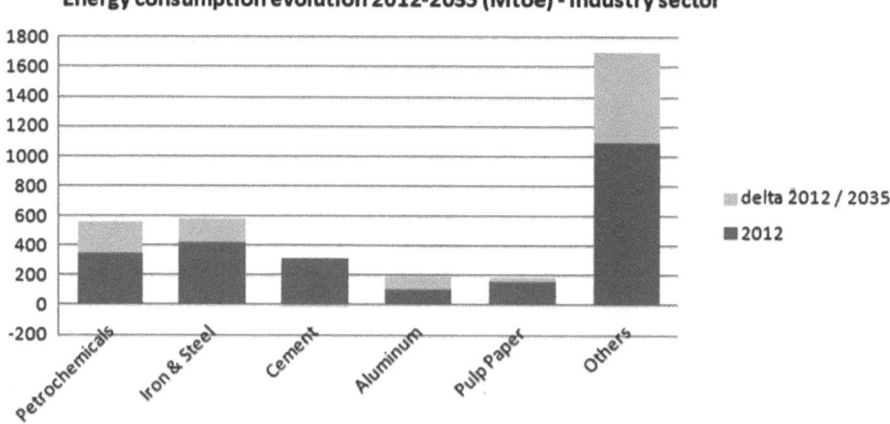

Fig. 5.2 Evolution of energy consumption in industry (© OECD/IEA, Statistics 2015; © OECD/IEA, WEO 2012)

savings. Most of the existing sites in OECD countries will not start revamping before 2040. However, the majority of industrial growth will be realized in new economies and by way of new plants being built, which therefore can be designed to incorporate new technologies right from the start. The production capacity in non-OECD countries shall be multiplied by four by 2050 (© OECD/IEA, OECD 2012; © OECD/IEA, Non-OECD 2012).

Important R&D has been devoted to rethink as well petrochemical processes with a view to optimizing them or to substitute oil. New production technologies using gas (less polluting than oil) and biomass could substitute classical naphtha steam cracking. Catalytic processing has already achieved yields of around 65% and engineers are working towards raising these to 80%. New separation methods, notably through the use of special membranes, could also lead to savings in the range of 5%, as the separation process accounts for around 45% of total energy consumption. Beyond these improvements, process intensification could lead to savings of up to 5% in the next decade, and probably up to 20% in the next 40 years. This intensification can be realized using different methods, including process miniaturization and thermodynamic improvement. Finally, better management of flows and return flows through improved time management can also lead to better efficiency of the overall process.

Other initiatives include replacing all plastics by new materials not based on primary fossil resources. Bioplastics could replace 80% of the world's plastics, according to experts (© OECD/IEA, Technology Industry 2009). Furthermore, better waste treatment and the use of new materials offer considerable potential savings.

Finally, petrochemical plants generate a significant amount of greenhouse gas emissions. CO_2 capture systems could be implemented to reduce the ecological impact of these plants.

5.2.1.2 Steel Industry

One of the largest energy consumers, the steel industry is a prolific emitter of greenhouse gases. According to the International Energy Agency (2009), up to 20% of the energy used in the steel industry could be saved.

In traditional steelmaking, a chemical reaction between coke (almost pure carbon) and oxygen inside a blast furnace creates carbon monoxide gas, which reduces and purifies iron oxides. This process is extremely energy-intensive, consuming around 14 GJ/ton. New technologies are more efficient. Electric arc furnaces, with an energy consumption of around 5 GJ/ton, bring considerable energy savings. However, the deployment of electric arc furnaces is limited, except in North America. Electric arc furnaces use scrap to produce steel so their use is therefore dictated by materials flow in the overall steelmaking process (Fig. 5.3).

Research is also focusing on new technologies for manufacturing steel in a more energy-efficient manner. Direct reduction of oxides for instance would not require any metal fusion. Other techniques using only electricity and electrolytic reaction are still being studied. Beyond developing new furnace technologies, substituting coke with a gas or the reuse of waste (like in petrochemical plants) could help reduce greenhouse gas emissions. Like in petrochemical plants, the reuse of heat generated to produce electricity is also an important area of energy efficiency.

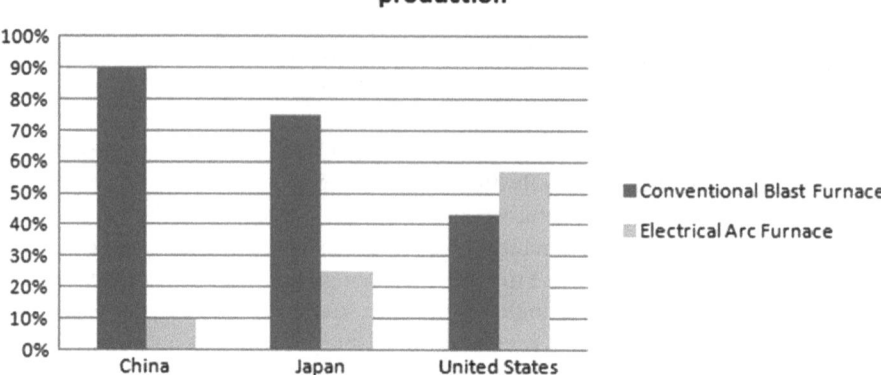

Fig. 5.3 Conventional blast and electrical arc furnace technologies (© OECD/IEA, Technology Industry 2009)

Finally, CO_2 capture systems are an interesting solution to reducing greenhouse gas emissions.

5.2.1.3 Cement Plants

Cement plant technologies have evolved a lot in recent years. Energy consumption is an important part of the direct variable costs of cement plants, and therefore a key factor of profitability. Recent technological developments have permitted operators to reduce energy consumption by around 25% (around 3 GJ/ton) (© OECD/IEA, Technology Industry 2009). The lifetime of cement plants averages 50 years, so the main challenge here is to renovate existing installations. As in other industries, the optimization of the heating process is key to improving energy performance. The introduction of preliminary heating could help reduce temperature gradients and therefore the energy required. The use of cooler grates instead of rotating coolers would also improve energy performance. Furthermore, reusing the heat dispersed to produce electricity (which could then be used back in the heating process) is an important opportunity to optimize energy consumption and waste. Using variable speed drives to adjust to the particularities of the process execution and the materials used would also help optimize in real-time the execution of the process. Finally, using waste or biomass as a replacement for fossil fuels is another way to reduce primary energy consumption. This is already widespread in OECD countries but has yet to be deployed elsewhere.

Finally, as in steel foundries, CO_2 capture systems are a possible option for reducing greenhouse gas emissions by cement plants.

5.2.1.4 Aluminum Plants

Aluminum plants consume enormous quantities of electricity. The deployment of the best technologies available would lead to a 13% reduction in energy consumption. Improving the process control is key to achieving this saving potential. Optimizing the chemical reaction can help reduce the energy required to achieve it. In addition, insulation work on the furnace can help better control the heat losses. The best plant technologies consume around 13.5 MWh/ton, which represents 13% savings compared to traditional installations (© OECD/IEA, Technology Industry 2009). The reaction which generates aluminum should theoretically not require more than 6 MWh/ton of energy. Today, research engineers are working towards getting closer to 11 MWh/ton by optimizing the traditional process from Hall-Hérault, which was designed at the end of the nineteenth century. Alternative methods are as well being studied, especially carbothermal reduction or kaolinite, both of which lead to theoretical energy savings of around 40%. These processes are however not industrialized as of today.

Greenhouse gas emissions, notably CO_2 production, can be limited using CO_2 capture systems. Finally, the aluminum manufacturing process being essentially powered by electricity, decarbonized electricity remains the primary way to limit greenhouse gas emissions.

5.2.1.5 Pulp Paper

The paper industry is extremely energy-intensive, even though in the end it only represents 4% of total energy consumption in the industry sector. Up to 20% of the energy used in the sector could be saved (© OECD/IEA, Technology Industry 2009). There are several ways to optimize energy performance in the paper industry. First of all, many improvements in the paper manufacturing process itself can be made to minimize energy consumption. For instance, the chemical elaboration of the paper pulp, the improvement of the efficiency of the motors which produce the paper sheets (using variable speed drives notably), or the optimization of the heating process all lead to significant energy savings. The most important source of savings comes from reusing the "black liquor", a waste product from paper manufacturing to produce electricity and heat, once it has been gasified. Pulp paper plants then become cogeneration power plants. The energy that can be retrieved from "black liquor" can make plants almost autonomous from an energy standpoint.

5.2.1.6 Electric Motor Systems

Electric motors are everywhere in industrial applications. They represent around 46% of total electricity consumption, or 7200 TWh/year. Around 64% of motors' electricity consumption happens in the industry sector. The buildings sector accounts for 33% and the remaining 3% is consumed in other sectors.

In the industry sector, electric motors' consumption tops 69% of total electricity consumption. Electric motors are the main source of savings in the non-electro-intensive segments, which account for the biggest share of the energy consumption in the industry sector (© OECD/IEA, Motors 2011; © OECD/IEA, WEO 2012) (Fig. 5.4).

Electric motor systems are thus critical to optimizing energy efficiency, in particular within the industry sector.

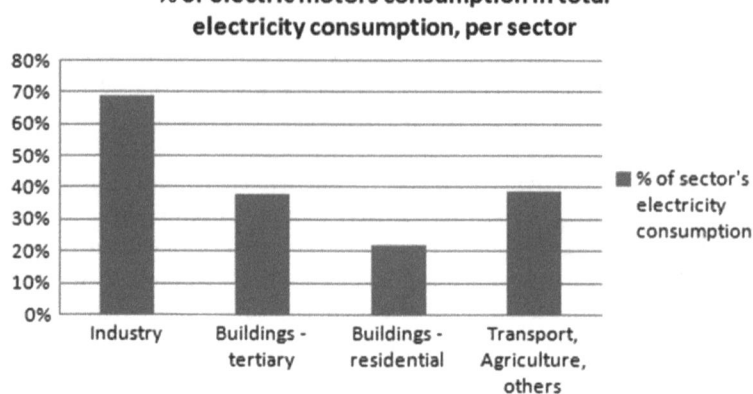

Fig. 5.4 Share of electric motors in total electricity consumption (© OECD/IEA, Motors 2011; © OECD/IEA, WEO 2012)

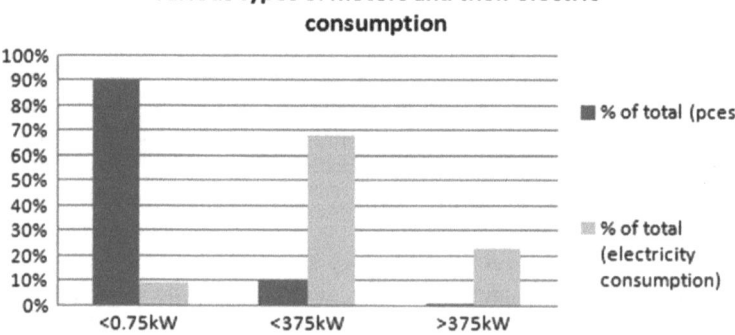

Fig. 5.5 Electric consumption of various motor types (© OECD/IEA, Motors 2011; © OECD/IEA, WEO 2012)

Around two billion electric motors are in use worldwide. Electric motors are highly standardized and have different power ratings. Motors with rated power below 0.75 kW have the largest share but consume only 9% of all electricity used by electric motors. The most electricity-intensive group of motors rates below 375 kW (and above 0.75 kW). There are around 230 million such motors in the world and they consume 68% of the total electricity consumed by motors, or around 31% of total worldwide electricity consumption. Finally, the largest motors (with rated power above 375 kW), consume 23% of total electricity of motors; 600,000 are in use (© OECD/IEA, Motors 2011) (Fig. 5.5).

Electric motors are used in four main applications: compressors (32% of electricity consumption), mechanical movement (30%), pumps (19%) and fans (19%).

To understand the efficiency of electricity consumption of a motor, it is important to consider the full system the motor is part of. Electric motor systems are made of the electricity supply system, the motor, and the mechanical equipment associated (transmission, gears, flow reducers, etc.). Electric motor systems can be extremely inefficient from an electricity consumption standpoint. According to the International Energy Agency (2011), 21% of electricity consumption could be saved between a traditional system and an energy-efficient system. The yield of the electric motor system could rise to 63%, compared to 42% for standard applications.

An energy efficient electric motor system is first made of high performance components. Mechanical parts can be selected with high-performance efficiency. For instance, transmission and gears can be selected accordingly. As an example, worm gears have high losses, other systems present better efficiency. Traditional transmission V-belts also have high friction when flat belts have a better yield. The efficiency of electric motors has considerably improved in the last few decades. High-performance motors (IEC60034/IE3 class) operate at around 90% efficiency for loads that vary between 50 and 125% of the rated power; while standard motors may yield as low as 50% efficiency when power is low. The quality of electricity also can impact the overall efficiency of the motor's operation as the electricity supply produces the magnetic field that helps convert electrical power into mechanical power.

An energy-efficient electric motor system has thus to be first designed for efficiency. This means avoiding intermediary parts (gears, flow reducers, etc.), and designing the motor system to the actual nominal operation of the process. In many applications though, it will remain impossible to design exactly such a system since the speed and torque requirements vary over time. This is the case for applications such as fans, pumps, escalators, and cranes. In these applications, the use of variable frequency drives helps tune the electric power that is supplied to the motors. This can yield an additional efficiency as high as 30%. Variable frequency drives are thus critical to raising process and energy efficiency.

5.2.1.7 Summary

The industry sector consumes around 2400 Mtoe of energy a year, a third of world energy consumption worldwide.

A significant share of this consumption comes from electro-intensive industries (petrochemical, iron & steel, cement industries, etc.). These industries also greatly contribute to greenhouse gas emissions. In each of those industries, a multitude of innovations or technological improvements can help improve energy performance. The vast majority of plants in electro-intensive industries use large quantities of energy to produce the heat needed to trigger chemical transformation or change of state. Heat is greatly wasted as it is released to the atmosphere once used. As the heating process uses fossil fuels, its production leads to greenhouse gas emissions. Innovations in the sector therefore first aim to optimize the heating process by better thermal isolation, automated control, or the use of new components or elements that help accelerate or intensify the process. Reusing the produced heat would also help increase the yield of the installations, either to produce electricity or simply to improve the yield of the heating process. Another way to lower energy expenditure is to replacing current processes with alternatives—still being researched—that require lower quantities of heat.

In non-electro-intensive industries, electric motor systems are the primary source of energy use. Electric motors account for 69% of the total electricity consumption in the industry sector. They can be significantly optimized through the deployment of high-efficiency motors, efficiency-based design and the use of variable frequency drives.

Finally, carbon capture systems, although expensive, can help limit the emissions of CO_2 to the atmosphere (Suez 2014).

All things considered, 25% of the final energy consumption could be saved in the industry sector.

5.2.2 Energy Efficiency Potential in Buildings Sector

The buildings sector is split between the residential and the tertiary (commercial and industrial buildings) segments. The residential segment corresponds to 75% of the total energy consumption by the sector. Given this, energy use improvements in the segment would bring more balance in energy use by the residential and the tertiary segments in the next 20 years (© OECD/IEA, WEO 2012). While the residential segment's energy consumption will grow by 35%, the energy consumption in the tertiary segment is expected to grow by almost 80% according to the International Energy Agency (2012). Both segments' energy consumption must thus be tackled (Fig. 5.6).

The usage of energy in those segments must also be evaluated as well as its evolutions in the next two decades. Heating (space and water) accounts for 50% of the total energy consumption in buildings, while appliances and cooking account for 20% each. Within 20 years, the International Energy Agency (2012) forecasts that 75% of the energy consumption increase will come from heating (space and water) and an increased number of appliances (refrigerators, washing machines, TV sets, connected equipment, etc.) (Fig. 5.7).

There are several solutions to improve the energy efficiency of a building. Most new constructions are naturally more economical from an energy standpoint. The main challenge with buildings is essentially the existing base, which sometimes dates back several decades, with extremely poor energy performances. By 2050, 75% of the buildings that exist today will still be there. The challenge to raise energy efficiency in buildings is thus tied to renovation.

Buildings can gain efficiency in three main domains: insulation, heating, and appliances.

5.2.2.1 Insulation of Buildings

The renovation of an old building leads to savings that can amount up to 75% (20% when the renovation is light). According to experts, insulation works can lead to savings of up to 60% of the energy used for space heating (© OECD/IEA,

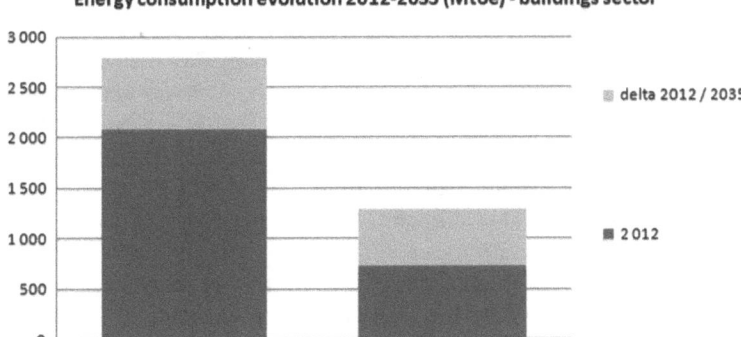

Fig. 5.6 Evolution of energy consumption in buildings (© OECD/IEA, Statistics 2015; © OECD/IEA, WEO 2012)

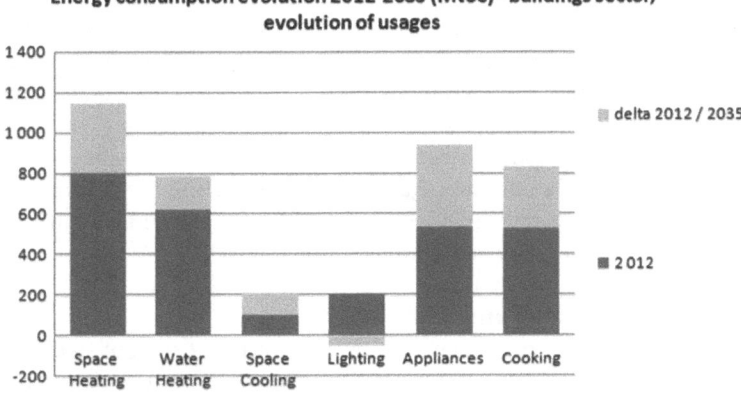

Fig. 5.7 Evolution of energy usages in buildings (© OECD/IEA, Buildings 2013; © OECD/IEA, WEO 2012)

Buildings 2013). The "Passivhaus" concept launched in Germany in the 1990s led to new buildings reaching insulation performances of around 15 kWh/m^2/year, which represented at the time 90% of savings in heating (Passivhaus 2014). Thermal insulation of a building starts with windows. It is generally admitted that the quality of windows can have an impact of around 10% on the total heating of a building (© OECD/IEA, Buildings 2013; © OECD/IEA, Efficient Buildings 2013). The efficiency of a window is measured using thermal transmission, with W/m^{2}K as the unit of measure. Single-glazed windows offer an average of 5.6 W/m^{2}K, against 1.8 W/m^{2}K for double-glazed windows. Modern technologies today integrate thin plates of aluminum to reduce heat transmission and reflect sunlight. Such windows have transmission levels of around 1.1 W/m^{2}K. In North European countries, triple-glazed windows have been introduced but they failed to take off in the market. To reach even lower transmission levels, the Japanese industry has designed vacuum glazing, which results in transmission levels of around 0.6 W/m^2 K. Finally, new dynamic glazing technologies adapt the transmission level to sunlight. Thermal insulation is also about walls. Without any insulation, the thermal transmission of a wall is between 3 and 10 W/m^{2}K. When properly insulated, transmission levels for walls do not exceed 0.5 W/m^{2}K. There exists different sorts of wall insulation, which can sometimes be combined. The first level of insulation is however the most efficient. Savings are not substantial beyond the first level of insulation. Insulation can be internal or external to the wall. Roofs are also an important source of thermal inefficiency. Sloping roofs are favored in cold countries because of snowfall, but they are more difficult to insulate. In hot countries flat roofs painted white reflect up to 80% of sunlight and are thus very efficient. Thermal insulation can be improved by improving internal ventilation. Ventilation is all the more efficient as the leakage rate is low. This rate is measured by putting the entire ventilation system under pressure and by measuring the number of air renewals within a given period of time—the higher the leakage rate, the more the number of air renewals. Leakage rate has of course consequences on space heating, which consequently becomes more important. The reduction of air flows between the outside and the inside leads to savings of between 5 and 40% of total heat consumption, depending on the ventilation system. Finally, the design of buildings and their envelopes also affect thermal efficiency. Windows and openings orientation and sun reflection systems are other options for improving the energy efficiency of a building. According to the International Energy Agency (2013), the simple use of reflective technologies in cities would lead to a temperature drop of two to four degrees within downtown areas.

5.2.2.2 Heating in Buildings

The second axis of energy efficiency improvement is the heating of space and water in a building. Here, there are significant differences between OECD countries and other regions. In new economies, biomass (waste, wood) remains the primary source of heating, accounting for around 70% of total heating in India and 50% in China, against less than 10% in Europe. In OECD countries, boilers are the main source of heating. The energy efficiency yield of traditional boilers is poor, around 70%. Electrical heating is much better, with a yield of 100%, but this needs to be balanced as electricity production is extremely inefficient, with yields around 30–40%. Boiler technologies have improved over the last years, and new condensation boilers can reach up to 95% yield. Going further, heat pumps have excellent yields of above 100% (which means the energy delivered is above the one necessary to its operation). Hybrid heat pumps combine several heating sources to limit the amount of energy necessary to deliver the desired temperature gradient. Hybrid pumps can for instance be coupled to renewable energy such as a solar or a geothermal energy source. The heating of water is also a significant area of energy consumption in a building, in particular in the residential segment. Traditional water boilers are inefficient from an energy standpoint as they provide hot water at any time of the day. Hot water storage can lead to energy inefficiencies of 10–30%, depending on the geographical region. The efficiency of water boilers can first be greatly improved by not storing hot water. Instant heating technologies significantly reduce the energy necessary for heating. Japanese industry has created a boiler which is powered by a heat pump. Half a million of such boilers are manufactured every year, but only in Japan. Heating can also be thought out more globally, beyond the efficiency of the equipment used. The typical yield of a thermal power plant is around 30–40%. The rest of the energy is generally lost as heat. Cogeneration reuses this heat and distributes it through a heating network. Consequently the yield of cogeneration plants goes up to 80 or 90% in the best cases. The main issue here is the deployment of cogeneration technology. Heating infrastructure need to be built or upgraded for this use. Furthermore, cogeneration plants need to operate at a specific nominal rate, or their efficiency decreases rapidly. The quantity of both electricity and heat also needs to be calculated precisely. Cogeneration is mainly used in North Europe; 40% of the electricity production in Denmark is done using cogeneration power plants.

5.2.2.3 Appliances in Buildings

The third way to optimize buildings' energy efficiency is to optimize the use of lighting and of the equipment connected. This is a key topic when looking at the actual penetration rate of appliances in residential buildings and their adoption in various regions of the world (Fig. 5.8).

Ownership of refrigerator is widespread across the world but that for washing machines is not. The rise of living standards in new economies as well as urbanization will lead to an increase of these penetration rates. Appliances' efficiency is thus key to total energy consumption optimization in buildings. Already efficiency has made strong progress over the last years. Sealing improvements, optimization of the use of heated water (via instant heating or heat pump technologies) have led to

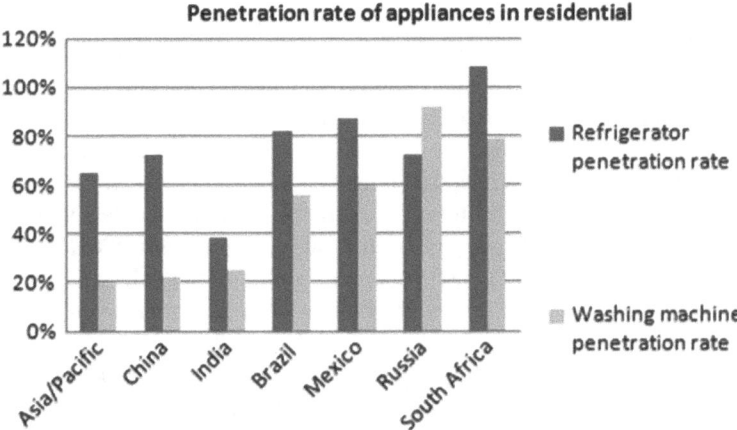

Fig. 5.8 Penetration rate of appliances in residential (© OECD/IEA, Buildings 2013; © OECD/IEA, Efficient Buildings 2013)

significant savings. Refrigerators' electricity consumption has dropped by 30% since 10 years ago and now stands at around 300 kWh/year. In some countries (e.g. Japan) the consumption was halved (© OECD/IEA, Buildings 2013). Like refrigerators and washing machines, TV sets and computers have become more prevalent and more energy-intensive. Improvements such as screen technology and standby capability have helped. Lighting is another major energy consumer (© OECD/IEA, Energy Efficiency 2013). It represented in 2010 200 Mtoe of energy. LEDs (light emitting diodes) will progressively replace traditional lamps and thus reduce significantly this number (© OECD/IEA, Buildings 2013; © OECD/IEA, Efficient Buildings 2013).

5.2.2.4 Towards Smarter Buildings and Homes

Besides these improvements, energy management technologies can strongly contribute to optimizing energy use in buildings. Energy management "active controls" are designed to operate real-time and as close as possible to user requirements to lower energy waste. They usually operate on lighting, heating and air conditioning, but often also accommodate other services such as security, video surveillance, access control and fire detection. As an example, "active controls" will tie the lighting of a room to occupancy and daylight level, preventing lights from remaining on if daylight is sufficient or if the room is unoccupied. These controls are called "active" since they operate in real time and automatically, as opposed to "passive" solutions such as insulation works, heating and appliance efficiency.

Savings from such controls will obviously vary with the type of building and its use. Some buildings have spaces with different usage (offices, hotels, homes, etc.), while others are more homogeneous (retail, logistics, restaurants, etc.). The more fragmented the building, the more opportunity for savings. Space fragmentation is a key factor of energy waste. Also, some buildings are used continuously (hospitals, data centers, etc.), while others are used intermittently (hotels, sport facilities, theatres, etc.). The more intermittent the operation, the more the opportunity for savings. Time fragmentation is another key factor of energy waste.

The efficiency of the control will depend on the granularity of space and time monitoring. For example, an energy management system in a building can consider the entire building as one zone, or split the building into multiple zones (staircase, offices, corridors, facilities, etc.), or control individual rooms as zones. Up to 60% of energy can be saved when "active controls" operate at the lowest granularity level both in terms of space and time fragmentation. Time granularity alone offers the highest efficiency independently from space fragmentation. With no split per zone, time granularity leads up to 35% energy savings (Schneider Electric 2014). The higher the granularity of control, the more precise the "active control", and the more the energy savings.

Schneider Electric headquarters, located in Paris, France, is a very good example of such type of "active" building. The building was the first in the world to receive ISO50001 certification. It consumes only 80 kWh/m^2/year, four times less than in the previous headquarters of the group, using an advanced building management technology (Schneider Electric 2014). Such building management systems are traditionally deployed in large tertiary buildings (commercial centers, offices, hotels, etc.), where it is easier to control energy efficiency. One of the main challenges faced by the industry is deploying similar technologies to small and mid-sized buildings (such as small bank agencies or shops) which have simpler needs and represent a much wider physical and energy footprint. In the residential segment, new information and communication technologies (ICTs) allow the user to measure in real time his energy consumption per main item (heating, cooking, lighting, etc.) and to optimize the use of his energy (regulating temperature, turning on/off equipment) by remote control over the Internet (Schneider Electric 2014).

"Active controls" deployment is progressing rapidly, particularly in new constructions. The technology is equally applicable to existing buildings, where it can lead to important energy savings. Payback calculations show that investments in "active controls" break even within 4 to 5 years, faster than "passive" solutions which payback is rarely reached before 10 years. This is due to the cost of installation. Typical insulation works cost between 30 (internal insulation) and 300 Eur/m^2 (window/double glazing) for energy savings in the range of 15%. "Active controls" cost from 20 Eur/m^2 (simple solution in residences) to 50 Eur/m^2 (room control and monitoring in tertiary buildings) for benefits ranging from 10 to 60% (Schneider Electric 2014).

Smart building and smart home solutions thus form a cornerstone of energy efficiency in the buildings sector.

5.2.2.5 Energy-Efficient Data Centers

Data centers are major energy consumers. Their energy consumption has been increasing by double-digit figures every year to power up the massive deployment of ICTs. There are several ways to optimize energy efficiency in data centers (FEMP 2011).

First, IT systems themselves can be optimized. Their consumption traditionally accounts for 50% of the total electricity bill, the rest being consumed by cooling and secured power. Servers are an intuitive first target for energy optimization. They run partial loads most of the time and the use of variable speed fans or power management devices can help optimize the power consumed by servers. Distributing the overall computing load across multiple servers (and processors) can also yield energy savings. Here, beneficial hardware and software technologies include multi-core processors and server virtualization. As well, storage devices can be optimized by managing accurately the data that needs to remain online against the data that can be stored offline. Network equipment has also considerably improved from an energy consumption standpoint. Well-designed network systems help reduce the energy footprint too. Finally, new high-efficiency power supplies have today an efficiency of 95%, against 70% for outdated technologies.

Air management and cooling is another area for improvement. Well-designed air flow through cold/hot aisles in data centers is key to limiting the heating of the various components and systems and the associated cooling required. Even proper cable management can help to improve air flow. Cooling must be well designed to ensure energy-efficient data center operation. There is a vast array of Computer

Room Air Conditioning (CRAC) systems available in the market. High-efficiency equipment that use variable frequency drives to adjust energy consumption to actual cooling need is favored. Central air handlers are also more efficient than modular systems (FEMP 2011). Equipment can also be cooled using direct liquid cooling. The liquid captures the heat generated by the equipment and drives it outside of the server room, instead of it being dispersed inside the room's atmosphere (which would then require additional cooling). Free cooling can also be used, with air-side or water-side economizers, which basically use the temperature gradient between the server room and the outside.

The power supply plays also a critical role in achieving energy efficiency. Load factor is a key element when it comes to redundant systems. Smaller uninterruptable power supply (UPS) units are preferred to larger ones as they increase the relative load factor and thus energy efficiency. The consolidation of redundancies (one power supply source per server rack instead of one per server) also helps better distribute power to the loads. Finally, DC power, required by many components, can be distributed in an organized manner in order to avoid multiple AC/DC conversions throughout the system, and the corresponding losses.

The heat generated by servers can also be reused to maximize the energy efficiency of a data center. Cogeneration can contribute to a better efficiency ratio. Wasted heat can also serve to keep standby generators warmed up, or to run absorption chillers as a complement to electrical systems.

Finally, the consolidation of data processing remains the best way to reduce energy use and its associated costs. Cloud-based solutions and collocation data centers are more efficient than small systems owned and operated by individual businesses.

5.2.2.6 Summary

The buildings sector in 2010 consumed 2800 Mtoe of energy, with residential buildings accounting for three quarters of the total. According to the International Energy Agency (2012), up to 20% of the energy used in buildings could be saved.

The main challenge for this sector is buildings' renovation. By 2050, 75% of the existing buildings will still be standing. Replacement rate remains very low. If not renovated, their energy consumption will continue to be much higher than that of modern buildings, which are more energy-efficient. The number of appliances used within buildings will increase dramatically in the coming 20 years, with corresponding energy consumption set to almost double within the same time period.

The first way to raise energy efficiency in a building is to improve its thermal insulation, through walls, windows and rooftops. The second way consists of optimizing space heating. This can be done by optimizing equipment efficiency, as well as by introducing renewable heating in the process. The third way is to optimize appliances' consumption and their usage. Finally, a large portion of the saving opportunity in the buildings sector will come simply from the better usage of energy. One of the main issues in energy use in buildings relates to how the space is used. Buildings are fragmented into floors, zones, rooms, with occupancy typically changing over the course of a day. Active controls help optimize the overall energy consumption by adjusting the energy needs of each "fragment" of the building depending on the time and the occupancy of the space. Such dynamic control could, according to experts, help achieve up to 60% energy savings in buildings (Schneider Electric 2014).

5.2.3   Energy Efficiency Potential in Transportation Sector

The main energy consumers in the transportation sector are light road vehicles (cars, buses, etc.). They represent half of the total energy consumption of the sector, and will correspond, according to the International Energy Agency (2012), to 40% of the total increase in energy consumption by 2035. Next comes air transportation. While the segment currently accounts for only 11% of total consumption, it will take a 40% share of the increase in energy consumption in the next 20 years, on par with light road vehicles. Consumption optimization in the sector will thus mainly be related to the optimization of energy use by light road vehicles and limiting the energy consumed by air transportation (Fig. 5.9).

5.2.3.1 Light Road Vehicles (Individual Transportation)
The most challenging issue concerns gasoline consumption by individual vehicles. There are today around 700 million vehicles in the world. Economic development could push the vehicle population to between two to three billion cars by 2050. Extensive research effort is being spent on reducing the environmental impact and gasoline consumption of individual vehicles.

Vehicle motorization has considerably evolved in the past decades. An engine is designed to operate at a nominal speed and power, so the main issue related to engine efficiency is related to its operation. The engine indeed runs at various speed

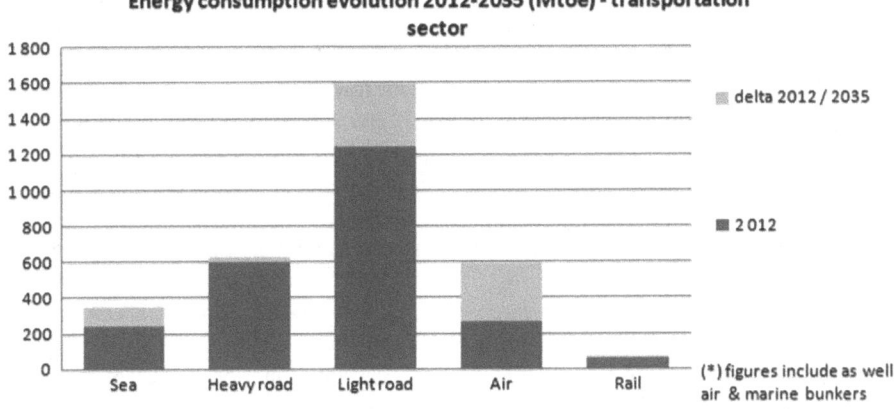

Fig. 5.9 Evolution of energy consumption in transportation (© OECD/IEA, Statistics 2015; © OECD/IEA, WEO 2012)

and power, depending on the conditions of use. It is rarely used at its optimal ratio. The design of motors has therefore evolved in the past decades to optimize the gas consumption, favoring its efficiency with various levels of speed and power. The challenge lies in the compromises that are being made to optimize its efficiency. Modern engines are better designed and tuned to limit gas consumption. Direct injection has progressively replaced carburetors, and engines have been made smaller capacity-wise, with turbochargers compensating for the reduced horse-power. All these have helped reduce gasoline consumption.

A recent motorization innovation, hybrid engines, combine gasoline and electric motorization and can deliver up to 30% fuel saving. Electricity production is done using regenerative braking—the rotating energy of the wheels is transformed to electricity by the same process as the one used in power plants. The electricity is then stored into a battery that can then serve as a generator. While they are available in the market, fully electric vehicles have yet to take off. There are various reasons for this. First, current batteries limit the travel distance of such cars. In addition, the infrastructure for battery recharging is far less widespread that that for gasoline vehicles. And while electric vehicles emit no greenhouse gases, they are no greener than conventional vehicles if the electricity used to charge their batteries is produced by conventional thermal power plants. Hydrogen motorization is another technology that has the potential to serve as a substitute for gasoline. Hydrogen is mixed to oxygen in fuel cells to create a similar reaction as the one happening in a conventional electrochemical battery. Hydrogen motorization only produces vapor. However, the cost of this technology is extremely high and the lack of refueling infrastructure has so far limited its adoption.

Beyond motorization, some more recent technologies also help increase vehicles' energy efficiency. The vast majority of transportation usages concerns small distances inside cities. In such travel, cars are often stopped. "Start/stop" technology automatically stops the engine when the car is forced to stay put for some time, and restarts it automatically when the driver depresses the accelerator. This leads to a sharp decrease of gas consumption in cities (around 7% on average). Another innovation further, valve rhythm adaptation, can also help improve the gas consumption of the engine by around 10%.

Transmission systems also play an important role in energy optimization. Maximum efficiency is reached with manual transmission systems. Automatic transmissions do not exceed 85% of yield, versus 97% of yield for fully manual transmissions. Now, as automatic transmission systems take a bigger market share, "manual automatic" systems have also been introduced, notably for small vehicles. They are essentially manual transmission systems which change the gears automatically, and give 90–95% of yield.

Aerodynamics also helps reduce gasoline consumption. Lightening vehicles by substituting traditional materials with aluminum and carbon leads to a gasoline consumption reduction of around 6% (for 10% weight reduction). The optimization of electricity consumption inside the vehicle can lead to 3% gasoline consumption reduction as well. Finally, tire manufacturing techniques also have an impact on gasoline consumption. New technologies reduce the friction on the road without compromising adherence, giving up to 5% reduction.

The final area in energy optimization in vehicles is the substitution of gasoline by biofuels. The two main biofuel producers in the world are the United States and Brazil. In most countries, gasoline can be blended with biofuels. Often 10% of the gasoline purchased is mixed with biofuels. In some countries like Brazil, this ratio is higher. There, most cars can run on only biofuels, and gasoline is generally blended with biofuels up to 25%. Biofuel production can be done using beans such as corn, sorghum and, like in Brazil, cane sugar. Cane sugar results in biofuels that are competitive against traditional gasoline. Other types of biofuels can be more expensive than gasoline. While biofuels help to reduce oil dependency, their use does not help to reduce greenhouse gas emissions—biofuel production does emit greenhouses gases. The agriculture of the crops that go into biofuels also generates greenhouse gases. According to Parmentier (2009), one ton of oil equivalent would be required to produce three tons of biofuels. Much current research focuses on reducing greenhouse gas emissions from the production of biofuels.

In summary, there are already a number of technologies which help to radically improve fuel efficiency. Obviously, vehicles that incorporate such technologies cost more. The extra cost can vary from a few thousand euros to 25,000 euros in the case of hydrogen or electrical motorization (© OECD/IEA, Transport 2009). The average lifetime of a car usually exceeds 15 years. There are today around 700 million cars in the world; 50 million are purchased new every year. This means that no more than 7% of the car total volume is renewed every year. In addition, this 7% combines both the replacement of old cars as well as the natural growth of the market. The actual replacement rate is therefore very low and market inertia is strong. Renewing entirely the world's fleet of cars will probably take several decades.

5.2.3.2 Heavy Road Vehicles (Trucks)

It is generally accepted that truck transportation is related to economic growth. It has also been proven (© OECD/IEA, Transport 2009) that the more GDP increases, the less the correlation. Indeed, while people get wealthier, they tend to have a lesser need for cumbersome equipment while their need for advanced and high-value-added products rises. Road transportation is thus naturally impacted.

The trucks currently in use already have a good efficiency, even though improvements can still be made. The thermal efficiency of truck engines is around 40% today. The International Energy Agency (2009) considers it could be brought up to 50 or 55% in some cases. As with cars, hybrid motorization also brings significant improvements to trucks, even though most of the trucks are used outside cities. Aerodynamics plays an important role, as well as auxiliary services on board. Tire performance is also crucial to reducing gas consumption; new high-performance tires lead to fuel savings of 4–8%.

Studies have also shown that increasing the size and weight of trucks lead to an overall saving of the gas consumed. Trucks that are 25 m long (60 tons of load) consume indeed 15% less fuel than the equivalent load delivered using 16-m-long trucks (40 tons of load). However, large trucks require a number of adaptations to road infrastructures.

Finally, the most important innovation for reducing energy use by trucks lies probably in the modernization of the logistics system itself. The logistics chain can be restructured by partially integrating production or by limiting the variety of supplies. This would considerably reduce the volume of exchanged materials and therefore the truck transportation volume. In addition, new procurement strategies which would extend, for instance, the duration of the deliveries would allow for optimal flow of supplies and therefore reduce the volume transported as well as empty load transportation.

5.2.3.3 Air Transportation

Air transportation in 2012 used around 260 Mtoe of energy, or 11% of the total consumption by the transportation sector. According to the International Energy Agency (2014), the figure should rise up to 600 Mtoe by 2035 and represent 18% of total consumption. Air transportation is expanding fast. OECD countries represented in 2005 two thirds of world air transportation; they should correspond to one third in 2050, to the benefit of the rest of the world, particularly Asia (© OECD/IEA, Transport 2009).

Energy consumption is at the heart of the productivity of this segment. The energy consumption of the new Boeing 787 is around 1.3 MJ/seat/km. This needs to be compared to the one of the Boeing 767 which it replaces, which is around 1.9 MJ/seat/km. This represents an annual saving of around 7 million liters of fuel, or 6.4 million dollars. Over the lifetime of the plane (around 30 years), this corresponds to significant savings (© OECD/IEA, Transport 2009). The increase of energy efficiency in air transportation thus has a direct impact on the segment's profitability.

Over the last 30 years, the consumption of fuel has been reduced on average by 35% (and 75% of the noise). Still, a lot remains to be done.

According to experts, propulsion systems could be optimized by another 15–20%. While innovations centered on motorization are essential, improved aerodynamics and reducing the weight of airplanes (using new carbon fiber materials for instance) bring the most benefits (around 20–30% less energy).

Changing the way air transportation is managed could also lead to 7–10% of additional energy savings. The improvement of air traffic control systems could help speed up the landing time of airplanes, in particular adopting a continuous descent slope (against a step-down today), which would reduce fuel consumption. Large "hubs" would also be helpful, by creating less direct routes and more flight connections. Profitability of air transportation is very much related to passenger occupancy and the "hubs" are essentially designed to maximize airplanes' occupancy. Any development against the use of "hubs" would hamper the development capability of airline companies and is thus very unlikely. Finally, the optimal distance in terms of fuel consumption is around 5000 km (© OECD/IEA, Transport 2009). This distance is calculated between the necessary overconsumption at the time of the takeoff and landing, and the necessity to carry important quantities of fuel over a long route. The reduction of the number of long-haul flights, split into a multitude of 5000 km routes, could lead to an overall fuel consumption reduction for the air transportation segment.

5.2.3.4 Marine Transportation

Marine transportation is a fragmented segment with a multitude of companies in many countries. This naturally limits innovation. Many options do exist to reduce the fuel consumption of ships. The right design can help lower fuel consumption. The size (4% of savings), the weight (7% of savings), and the ship's profile (5% of savings) help to reduce the fuel consumed on the ship. Motorization can help, too, notably by reusing the heat produced by the engines, with savings potentially reaching 10%. Propulsion systems offer also potential savings of around 10%. Finally, better management of energy distribution onboard the ships using automated process management could offer 10–15% of fuel savings. The duration of time a ship spends in the harbor also has an impact on the energy it consumes. Most ships need to keep their engines running at very low efficiency levels to power up their auxiliaries while at the dock. A direct supply of electricity from the harbor to the ships at dock is an essential solution to reduce drastically the greenhouse gas emissions and the fuel consumption.

5.2.3.5 Smarter Transportation

Beyond improvements that can be made to cars, trucks, planes and ships, changes in habits can also be a powerful vehicle of energy efficiency.

Eighty percent of transportation is for short-distance travel. This ratio even tops 94% in South East Asia. Short distances are thus the cornerstone of transportation efficiency. Optimizing short-distance travel will help reduce greenhouse gas emissions by the sector while improving mobility, in particular in new economies.

Short-distance travel concern mainly daily activities such as commuting (going to and from work), shopping or other personal activities. According to the United Kingdom Department for Transport (2009), 85% of transportation in England is devoted to these activities, which do not exceed 18 km/day. In the United States, these activities represent 91% of total transportation, with mileage below 20 km (US DoT 2009). Commuting alone represents around 30% of total travels. Individual vehicles used for these activities remain unused 90% of the time (CTA 2014).

Travel is highly dependent upon the context. First, rural areas are more difficult to serve with alternative transportation than cities, and distances are often longer, leading to a larger share of cars in total transportation use. In the United States for instance, 92% of travels are done by car in rural areas, against 78% in cities (US DoT 2009). This situation is much different in non-OECD countries, where rural areas still remain remote from large business centers. In China, the overall average distance travelled per year barely reaches 1000 km per individual, compared to 5000 km/year in Chinese cities (JTLU 2010). Chinese rural areas remain therefore extremely isolated. The context also varies a lot with the stage of economic development. OECD countries rely mostly on individual cars for transportation, with almost 60% of car use in Europe and close to 80% in the United

States. There are up to 600 cars for 1000 persons in the United States, and around 500 in most European countries, while only 40 in China and 15 in India (Road Transport 2012). In new economies, bus and rail prevail (around 60%). Most individual travels are done using two-wheelers, or non-motorized transportation (walking or cycling).

From an energy standpoint, transportation inefficiency can be described by the greenhouse gas emissions that an individual travel contributes to. The IPCC (2007) explains that individual cars are five times more polluting on average than bus or rail; two-wheelers are twice more polluting than bus or rail. This difference is essentially due to the load factor. Buses traditionally accommodate around 40 people in a single vehicle, while the average occupancy in individual cars rarely tops two passengers, and even 1.5 on average for commuting, the largest share of short-distance travels (UK DoT 2009).

Transportation efficiency will thus mainly be achieved by limiting the footprint of cars. In OECD countries, this corresponds to a reduction of the share of cars in short-distance travel, notably for commuting. In non-OECD countries, this will rather correspond to preventing this share from growing beyond a certain limit. In China, the number of cars could top 100 million in the coming 10–15 years (JTLU 2010).

The International Energy Agency (2009) lists several directions for reducing the share of cars in short-distance transportation. The first set of solutions consists of constraining their use inside cities and within suburbs. Toll parking and road pricing are solutions that can help limit the circulation of cars efficiently, as proven in several large European cities.

A second set of solutions is to offer new mobility solutions that can serve as efficient alternatives to individual cars. Public transportation networks can be developed to offer better service with a travel time efficiency that can compete (if not supersede) car travel. Metros and trams are already well developed in large cities in Europe, but much less in other parts of the world. Brazil (and then the rest of Latin America) also developed Bus Rapid Transit (BRT) systems which help accelerate travel timing. Buses travel in specific lanes, with prioritization at intersections; they offer higher capacity than traditional buses, with rapid boarding systems. As a result of this success in Latin America, several cities in China have already deployed BRT systems (JTLU 2010). Going further, the development of non-motorized transportation can also be supported. Just over a third of inner-city travel in Amsterdam (in the Netherlands) is done using bicycles. Specific adjustments to the road infrastructure can be made to encourage the development of carbon-free transportation modes. In several cities in Europe, bicycles can already be rented and shared across the population. When cars need to be used, carpooling offers a new solution to limit the number of cars in circulation. Same as for bicycles, this consists of using a given pool of rented cars. The intent is to maximize the usage of individual cars in a city, considering cars are unused 90% of the time (CTA 2014). Car sharing is another solution. Here, the intent is to increase the load factor, the average occupancy rate per vehicle. In the United Kingdom or in the United States, studies have measured that the occupancy rate barely reaches 1.5

passengers per car for commuting (30% of short distance travels). Information and communication technologies strongly enable the development of such new collaborative solutions.

A third set of solutions looks at urban design and land planning. The design of cities is obviously strongly constrained in existing ones, but many new cities rapidly develop in new economies. There were only 10 cities of more than ten million people in 1990. This figure is now 28 and shall go up to 41 by 2030 (UN 2014). At the same time, the number of cities with more than 500,000 inhabitants will grow in the same proportion. There were 560 such cities in 1990. There are today more than 1000 of such cities, with 300 more expected by 2030. The urbanization rate worldwide is projected to reach 70% by 2050, with an additional 2.2 billion people living in cities (ESA 2014). City planners can help minimize transportation by designing road infrastructures which facilitate the use of public transportation or of non-motorized vehicles for small-distance travel. They can also organize the landscape of work offices, homes, shopping and recreation facilities such that there is minimal need to use transport. Several new cities in the world (Masdar in UAE, King Abdullah Economic City in Saudi Arabia, Songdo in Korea, Dongtan in China, etc.) are currently being developed following these principles.

Finally, transportation efficiency can also be achieved by simply doing away with travel. Remote work is a good example. Almost a third of short-distance travel is dedicated to commuting; remote work helps cancel the unnecessary travel. In Northern Europe, where remote work is most developed, above 10% of the population regularly works remotely (at least 1 day a month), and 40% of the population uses this possibility occasionally. This new opportunity is developing fast and expanding beyond the borders of this region. ICTs have also been offering for several years increased opportunities for e-business. In such transactions, the delivery of products is done by the supplier directly. This allows the supplier to group deliveries, therefore reducing overall transportation, since each consumer does not have to go to the mall anymore.

Many solutions thus exist to drastically reduce the use of individual cars in OECD countries and to avoid the dramatic rise of congested, polluted and inefficient cities in non-OECD countries.

5.2.3.6 Summary

The transportation sector (including air and marine bunkers) in 2010 consumed around 2400 Mtoe of energy (© OECD/IEA, WEO 2012). There is in this sector massive potential for energy efficiency. In the light road vehicles segment, the energy consumption is mostly by individual cars and two-wheelers (© OECD/IEA, Transport 2009). Modern motorization technologies offer considerable fuel savings versus older technologies. Besides motorization, aerodynamics and materials substitution to lighten weight (notably in maritime and air transportation) also offer potential for energy efficiency.

One main issue in energy efficiency in the transportation sector is the very low replacement rate of cars. As a result, the energy consumption of the sector keeps growing. Proactive actions to deploy motorization technologies would have a considerable impact on the energy performance of the sector, and help tame the energy consumption growth of the sector. Up to 31% of the total energy consumption in the transportation sector could be saved.

Finally, beyond vehicle technology deployment, smarter transportation is possible. The bulk of transportation inefficiency lies in short-distance travel, particularly daily commuting. Replacing short-distance transportation by cars with public transportation or non-motorized vehicles, developing car sharing/pooling, and supporting remote work and e-business are among the solutions to increase drastically the energy efficiency of this sector.

5.2.4   Regional Perspectives

Massive energy efficiency potential exists across sectors. The deployment of the best available technologies could lead to up to 25% of savings in the industry sector, 20% in the buildings sector, and up to 31% of savings in the transportation sector. Overall, up to 25% of end-use energy—21% of primary energy—could be saved. When considered together, this potential and the actual consumption per region and per sector yield a regional perspective of where energy efficiency could be best realized. The potential of energy savings in a given region is calculated based on the mix of each sector in the overall energy mix of the region (Fig. 5.10).

Regions with a high mix of buildings versus transportation or industry have a higher potential overall. The overall end-use energy efficiency potential is also based on the baseline of the energy consumption. North America, Europe, China and Asia/Pacific present the highest opportunities for energy savings.

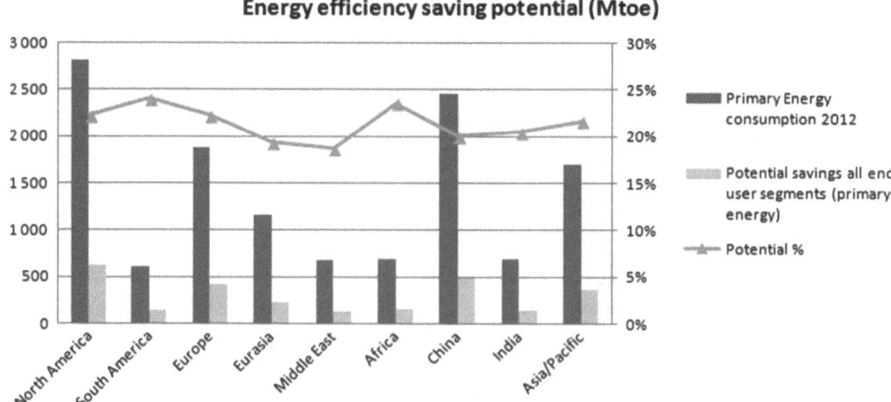

Fig. 5.10 Energy efficiency saving potential (© OECD/IEA, Buildings 2013; © OECD/IEA, Efficient Buildings 2013; © OECD/IEA, Energy Efficiency 2013; © OECD/IEA, Motors 2009; © OECD/IEA, Statistics 2015; © OECD/IEA, Technology Industry 2009; © OECD/IEA, WEO 2012)

5.3 Spectacular Waste of Electrical Energy

Almost three quarters of the 13 billion tons of oil equivalent of primary energy produced every year are actually transformed (© OECD/IEA, Explore 2014). The transformation of oil into various products (gasoline, plastics, pesticides, etc.) generates around 16% of wasted energy. The transformation to electricity generates around 60% of losses and is the major source of wasted primary energy resources. As a whole, around 30% of primary energy is wasted. A full 80% of that waste comes from electricity generation. Electricity production is thus extremely wasteful (Fig. 5.11).

Traditional thermal power plants have an efficiency of around 30–40%. This means that three units of primary energy are required to produce one unit of transportable electricity. Conventional electricity production is therefore incredibly inefficient. To produce electricity, an alternator transforms the rotating energy of a turbine into electricity through an electromagnetic reaction. Mechanical energy is therefore required to produce electricity. Most of the time, this energy comes from gas under pressure which, as it expands through the turbine, forces the rotating movement of the turbine. This gas can be obtained from any coolant. It can be steam or gas. In combined cycle plants, there are even two generators: one supplied with natural gas, the other with vapor generated from water heated up by the exhaust of the gas from the first generator. In any case, the gas needs to be heated up to expand and therefore create mechanical pressure on the turbine. It is this heating process (burning of oil, gas, coal or nuclear fission), in a boiler, followed by a double transformation into mechanical and then electrical energy which leads, by its natural inefficiency, to yields around 30–40%.

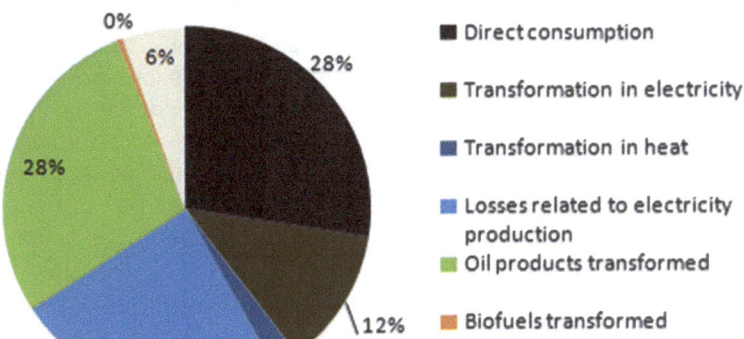

Transformation & Losses of primary energy

- Direct consumption
- Transformation in electricity
- Transformation in heat
- Losses related to electricity production
- Oil products transformed
- Biofuels transformed
- Losses related to oil transformation

Fig. 5.11 Transformation and losses of primary energy (© OECD/IEA, Explore 2014)

5.3.1 Optimizing Conventional Electricity Generation

The process can however be optimized. Richard Campbell (2013) analyzed the performance of coal power plants in the United States. According to him, the yield could be improved by an average of four points. While each plant is unique, the sources of inefficiencies are the same everywhere. They come from two main factors. First, mechanical imperfections in the various ventilation systems and pumps which circulate air and vapor through the system can deteriorate over the course of the power plant's lifetime. This deterioration leads to decreased performance of the mechanical conversion of the plant and, therefore, of the overall production of electricity achieved for a given quantity of fossil fuel. Leakages and lack of isolation lead as well to a lower efficiency of the heat transformation and to issues with maintaining the gas pressure at desired levels. Other than mechanical imperfections, the overall management of the power plant operation also contributes to inefficiency. A power plant is designed to operate optimally at a certain level. Operating a power plant at other than this level, whether below or above it, leads naturally to lower yield because of the resulting thermal inefficiency. Production variations can also lead to decreased thermal efficiency. A brutal drop of the electric power output followed by an increase requires raising the boiler temperature back up and therefore to increased consumption of fossil fuel; stable production makes better use of fossil fuels to maintain temperature and pressure levels. Finally, the combination of different variables (oxygen, temperature and pressure), if not optimized, decreases the overall efficiency of the process. Optimizing this combination requires the synchronized operation of different elements of the system: pumps, ventilation systems, burner, etc. Real-time process optimization is thus one of the primary sources of efficiency (or inefficiency) of thermal power plants.

Pilot projects have been run in the United States that use renewable energy such as thermal solar or biomass to complement the traditional heating process. This allows plant operators to minimize the amount of fossil fuel required by the heating process.

All these optimizations grant a few points of additional efficiency. Beyond this, the process itself needs to be revisited. Supercritical plants improve the yield of the Carnot cycle by increasing the pressure and the temperature of the vapor. This can result in yields of 46%, more than half more of the 30% yield for traditional plants. In such power plants, the pressure is up to 285 bars and the temperature around 620 degrees (Hansen and Percebois 2015). Circulating Fluidized Bed technologies operate at a lower burning temperature and less greenhouse gas is emitted while operating the boiler. Coal gasification creates a synthetic gas which does not generate CO_2 emissions. These technologies are mature but require heavy investments—they generally cost double what a traditional coal power plant does (Barré and Mérenne-Schoumaker 2011).

Gas power plants have a yield around 40%, slightly above that of coal power plants. Combined cycle power plants provide a much better yield, around 60%, as they use part of the heated gas that goes through the first turbine to heat up vapor in a second cycle which then will expand through a second turbine. This doubling of the production cycle leads to a higher power output for a given quantity of gas initially used.

Cogeneration plants (L'Expansion 2012) operate with two circuits. The first fluid is heated to operate through a turbine and cooled down while in contact with a second fluid. The second fluid, heated up by the thermal exchange, is then used in a district heating network. The total yield of a cogeneration plant can reach 80–90% in the best cases. This is a significant improvement over the yield of conventional plants. Now, for cogeneration to be efficient, the calibration of electricity production and heat must be well done. Any deviation from nominal operation leads to a sharp decrease of the power plant yield.

To sum up, there are a number of solutions to optimize the yield of a power plant. However, their yield is bound to remain extremely low, except when using combined cycle gas plants or cogeneration plants. This means that, whatever improvements are made (and they must be made), the level of losses remains colossal. Even though CO_2 emissions can be tamed to a certain extent with the use of carbon capture systems (© OECD/IEA, Coal 2014), such waste cannot continue to exist in a world where the demand for fossil fuels keeps increasing and where greenhouse gas emissions are skyrocketing. This is why engineers have long started to look for alternative energy sources, such as nuclear or renewable energies.

5.3.2 The Energy of the Stars

Nuclear energy is an alternative to fossil-based electricity generation. It does not consume fossil fuels but uranium, and does not generate any CO_2 emissions. Nuclear energy is deployed today in most OECD countries and expanding in other regions of the world, notably those with considerable needs in term of power, such as China or India. Consequently, the market growth is expected to range between 1.3 and 3.8%. It would be one of the highest growth rates among all energy sources.

Nuclear technology continues to evolve. Current nuclear power plants are of the third generation. The design has evolved but the main architecture remains the same; new innovations are being worked on. Fourth-generation reactors should come up around 2030. They will change drastically the energy equation. The current vast majority of reactors use Uranium 235 and Plutonium 239. The use of "slow neutrons" drives the fission reaction. Present-day nuclear fission produces "fast neutrons", which travel at around 20,000 km/s. A moderator is used to slow them down (Furfari 2007; Vendryes 2001). Research engineers have put together a new technology called "fast neutron" technology which slows down the neutrons to only 500 km/s. There are today three of such plants in the world. Two of them are in Russia and the third one is in Japan (Durand 2007). These reactors can be configured to produce more plutonium than what they consume, and are called nuclear breeders. With these reactors, Uranium 238, an isotope of uranium 140 times more abundant than Uranium 235, can be used for the fission reaction

(Durand 2007). As well, this technology helps produce 60 times more energy than a conventional third-generation plant with the same quantity of uranium. Uranium 233 can be used as well for the reaction. This isotope does not exist naturally but can be manufactured from thorium, which is four times more abundant than Uranium 238. Eventually, nuclear breeder technology could lead to around 300 times more electricity production capacity than current technologies, with the level of waste considerably reduced.

Going further, a new step in nuclear energy could also be reached. Nuclear fusion is a reaction where two atomic nucleuses assemble to form a heavier nucleus. A considerable amount of energy is generated from this reaction. This fusion is naturally at work in most of the universe's stars, particularly our sun. The hydrogen bomb was designed on this principle. The main issue in nuclear fusion is to control (and tame) the reaction in order to be able to produce energy continuously. The process uses deuterium and tritium, both hydrogen isotopes. Large quantities of deuterium can be found in seawater and tritium can be manufactured, so nuclear fusion fuel is potentially unlimited.

The nuclear fusion process requires bringing the reactor temperature to several million degrees. This technology exists today in Tokamak reactors (CEA 2014), but the reaction still cannot be controlled long enough for this type of reactor to be viable. Nuclear fusion could thus offer, provided it can be controlled, unlimited energy with very little waste. Of course the deployment of this technology would require considerable investment to build the corresponding power plants but it would solve definitely the electricity supply issue.

Like nuclear fission, nuclear fusion presents the advantage of offering considerable amounts of power, something renewable energy cannot do. However, it has the same limitation as nuclear fission. Considerable investments are required to build and operate the plants.

Other than nuclear energy, renewable energies are probably the most promising prospect of the electricity industry in the years to come.

5.3.3 Renewable Energies to Drive a New Paradigm

An alternative solution to the colossal waste from conventional electricity generation exists. Renewable energies are abundant, unlimited and free. They represent a complete change of paradigm. The fuel is free, available and permanently renewed. Consequently, once the investments are amortized, energy becomes almost free.

5.3.3.1 Huge Potential

Renewable energy reserves are a new concept that was developed a few years ago. It aims to measure the ultimate potential of use of renewable energy in a given place. The calculation considers first that nothing dramatic is to happen to current climatic conditions and therefore that the potential of use of renewable energy is to remain sustainable over time. Then, the calculation measures the reserves by looking at all areas where renewable energy sources could be developed, such as

Table 5.1 Renewable energy reserves (Favennec and Mathieu 2014; © OECD/IEA, WEO 2012)

Reserves (Mtoe)[a]	Consumption 2013	Theoretical available reserves	Expected consumption 2035	% of use in 2035 (%)
Hydro	914	2137	1474	68,966
Wind	103	322,810	420	0.130
Geothermal	6	10,760	95	0.880
Solar	30	5,595,381	172	0.003
Marine	0.65	12,284	6.14	0.050
Total	1053	5,943,373	2167	0.036

[a]Figures are built from TWh production converted in Mtoe, with a multiplier to account for the current yield of conventional plants. The purpose is to be able to compare renewable "reserves" to the current primary fossil fuel reserves

for instance the Sahara desert for sun power, or large continental shelves for wind. Every place suitable for the production of any kind of renewable energy is considered a reserve. Obviously the definition is highly theoretical. The evaluation should eventually include all specific issues related to land, neighborhood, etc. Nevertheless it shows the overall potential that renewable energy offers. The table below shows the production level in 2013, the theoretical possible reserves (Favennec and Mathieu 2014), as well as the estimated production level in 2035 (© OECD/IEA, WEO 2012), in million tons of oil equivalent. It shows that hydroelectricity would be well used, with a ratio in 2035 of around 68% of the ultimate potential. Other sources of renewable energy are however underused considerably. Solar is particularly underused with less than 0.003% of the ultimate potential consumed, despite the very strong forecasted growth of solar-based electricity production (Table 5.1).

In comparison, proven reserves of oil amount to 238,000 Mtoe. They correspond to two thirds of the wind potential reserves, and 4% of the sun potential reserves, which renew themselves! Renewable energy thus represents, theoretically, the solution to all energy problems of the planet, and forever. Solar energy in particular is by far the most promising technology.

5.3.3.2 The Photovoltaic Solar Potential

Photovoltaic solar electricity is probably the electricity production mode which offers one of the greatest opportunities. The ultimate potential (or energy reserve) of solar is indeed considerably higher than other forms of renewable energy. The sun transmits around 122,000 TW of energy every second to the Earth, out of which 88,000 TW are reflected back towards space. The rest is absorbed and contributes to the natural greenhouse effect. This massive amount of power corresponds to 3500 times the power that humanity would consume in 2050 (© OECD/IEA, Solar 2014). Favennec (2014) estimates that the ultimate amount of solar power that could be captured on the planet to produce electricity averages 26,000 TW. The current forecast of solar electricity production in 2035 (© OECD/IEA, WEO 2012), yet very conservative, corresponds to only 0.003% of this potential. Solar electricity thus represents a fantastic opportunity.

Solar energy is everywhere. Obviously, the global horizontal irradiance (GHI) varies across countries. It averages 1200 kWh/m^2/year in Europe and tops 2300 kWh/m^2/year in the Middle East. Every country receives sunlight so solar energy can contribute to energy independence.

Solar panels can be installed on the rooftop of every house in the world to produce electricity. This pervasiveness is unique to solar energy. With an average production of solar panels of 140 kWh/m^2/year, worldwide electricity demand would be met with 125,000 km^2 of solar panels, or 0.08% of the planet's total ground surface. As an example, for France, this corresponds to the rooftops of half of the buildings already existing (Manicore 2014).

Photovoltaic solar production has grown rapidly, with an average 49% growth in the last 10 years. In 2013, 37 GW of production were installed, bringing the total installed base to 135 GW. The overall level of investment reached 96 billion dollars. China was the top country in terms of photovoltaic solar installation, with over 11 GW installed during the year, followed by Japan and the United States (© OECD/IEA, Solar 2014).

5.3.3.3 Photovoltaic Solar Competitiveness

Despite this positive trend, the broad development of photovoltaic solar remained limited for years due to the high level of competition. The cost of solar is essentially based on the fixed initial costs. Operating and maintenance costs are very low and "fuel" is free. The cost of electricity production from solar is therefore made of the initial investment's cost as well as of the cost of financing, which depends on the weighted average cost of capital (WACC). Figure 5.12 shows the impact of WACC on the levelized cost of electricity (LCOE) (© OECD/IEA, Solar 2014). LCOE is the required price of electricity for investment to be profitable. It is based on a number of assumptions related to the investment's cost (electricity output of solar energy, cost of the system, etc.), but clearly reflects the importance of the cost of capital.

A typical WACC of 8% is considered in the solar business, and is at such considered in the International Energy Agency projections. Obviously WACC can vary, depending on the market situation and the specific context of the country where the solar business is being developed. It can be much lower; in countries like Germany, it is around 3–4% (Fraunhofer 2013).

One element which impacts the levelized cost of electricity is, of course, the amount of electricity that a solar system generates. For a given level of investment, increased performance of the solar system reduces the weight of the investment over the total cost (and consequently the weight of the cost of capital as well) (Fraunhofer 2013) (Fig. 5.13).

Another factor that contributes to build solar prices is the investment's cost. It is made of two main components: the cost of photovoltaic modules, the key active components that turn solar radiation into electricity; and the cost of the balance of system (BOS), which turns DC current produced by the modules into grid output. Figure 5.14 shows the cost breakdown of solar systems for a typical investment (Irena 2012).

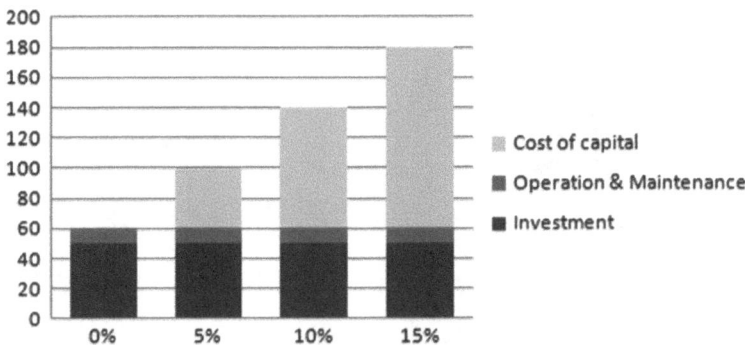

Fig. 5.12 Cost structure of photovoltaic solar (© OECD/IEA, Solar 2014)

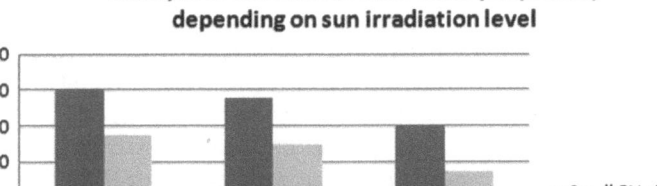

Fig. 5.13 Variation of solar LCOE with sun irradiation (Fraunhofer 2013)

The cost of photovoltaic modules represents typically between 50 and 60% of the total installation costs. The rest of the cost is balanced among the variety of activities and equipment needed to produce grid output.

The cost of modules have considerably reduced over the last few years. They were divided by five, down to around 0.8 USD/W, in 2013. The larger scale of production as well as the strong development of the industry in China contributed notably to drop these costs. The production of solar modules has moved from 5000 panels to around 35,000 panels between 2008 and 2013. Most of this growth came from China, which represented in 2013 over two thirds of global production. Modules have also become more efficient. Traditional c-Si modules based on purified silicon have reached efficiencies above 20% (© OECD/IEA, Solar 2014); thin film modules are catching up. Finally, concentrated photovoltaic technologies (CPV) used for satellites are the subject of many innovations, with

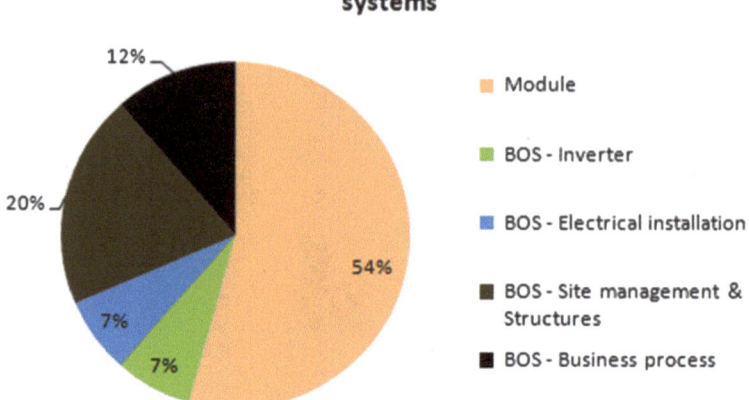

Typical Cost of investment of Photovoltaic Solar systems

Fig. 5.14 Cost of investment of photovoltaic solar (Irena 2012)

the ambition to reach levels of efficiency over 40% in the coming years (© OECD/ IEA, Solar 2014). The solar module market has thus become a global commodity market, with a permanent race to increase efficiency and price competitiveness. The International Energy Agency (2014) expects the cost of solar modules to further drop in the next 20 years, to range durably around 0.3–0.4 USD/W.

The other key element of cost for solar systems is that of the balance of system. Balance of system costs have also been divided by three in most mature markets. Systems' costs may differ greatly depending on the size. Utility-scale systems' costs are traditionally lower than those of individual house rooftop systems. The price variations essentially lie with the various business related costs, such as customer management, permits and grid connections. Total utility-scale systems' costs (including solar modules) reached 2 USD/W in 2013, while rooftop systems' costs ranged around 3–4 USD/W. These costs vary a lot, depending on the region. In Australia, rooftop systems' costs went down to 1.8 USD/W and utility-scale systems' costs were higher at 2 USD/W. In the United States, rooftop systems' costs reached in 2013 4.9 USD/W and utility-scale systems cost 3.3 USD/W. Despite the differences, all systems' costs went down everywhere by two-digit percentage drops. The International Energy Agency (2014) expects this trend to continue. Forecasts estimate that costs could go as low as 1 USD/W for rooftop systems and 0.7 USD/W for utility-scale systems in the coming 20 years. The simplification of grid connection procedures and business operations as well as the standardization of systems' architectures will help push down the costs. These cost projections do not include the additional cost of energy storage, which is a strong enabler of proper solar interconnection into the grid (© OECD/IEA, Solar 2014) (Fig. 5.15).

With the sharp decrease of photovoltaic solar costs, the corresponding cost of electricity production from solar has become more competitive. The LCOE has already decreased significantly and shall continue to drop. In 2013, it averaged around 200 USD/MWh, with strong variations depending on the country, the type

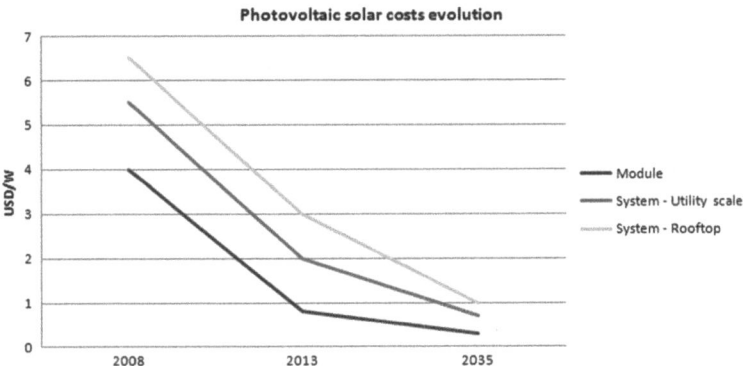

Fig. 5.15 Photovoltaic solar costs evolution (Fraunhofer 2013; © OECD/IEA, Solar 2014; Irena 2012)

of system (utility-scale or rooftop) and the technology used. Middle Eastern and African countries are favored due to the level of sun irradiation, which increases the performance of photovoltaic farms. Germany is favored over Spain thanks to a lower cost of financing. Utility-scale systems are less expensive than rooftop systems overall. Finally, solar panels manufactured in China are less expensive than those from other regions (the United States, Europe), due to a variety of reasons, particularly scale of production. The International Energy Agency (2014) forecasts average LCOE to go down further to an average of 70–95 USD/MWh by 2035.

The LCOE for utility-scale systems has to be compared to that for conventional technologies as both compete in the wholesale market. Lazard (2014) studied how unsubsidized renewable energies in the United States compete with other conventional technologies. Clearly, utility-scale solar systems are now beginning to compete with traditional conventional fossil fuel-based systems. What 5 years ago would have appeared unthinkable is now becoming a reality. Within the next 20 years, the additional competitiveness of solar systems will lead to a further drop in LCOE and a progressive substitution of traditional conventional plants (Fig. 5.16).

The true revolution, though, will take place on the rooftops of houses equipped with solar systems. The LCOE for rooftop systems needs to be compared to retail prices. When solar prices become close to end-user retail price limits, it makes more sense for the consumer to produce his own electricity using solar energy. The massive deployment of rooftop solar systems becomes then possible. The graph below maps the different countries' retail prices (© OECD/IEA, Electricity 2012) over a LCOE range for photovoltaic solar energy (WEC 2013). Solar levelized costs of electricity vary a lot depending on the year, the region, the technology installed and the cost of financing. The ranges below are a best approximation for 2013. The graph shows that solar competitiveness is already a reality in several

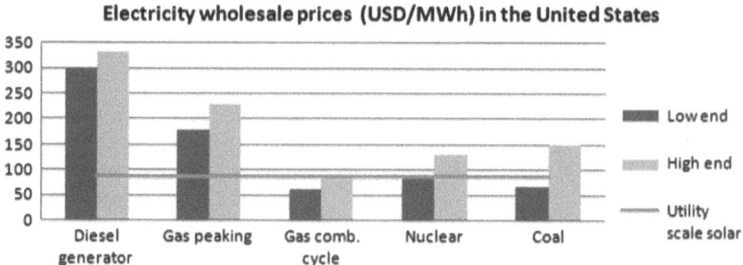

Fig. 5.16 Utility scale solar competitiveness (Lazard 2014)

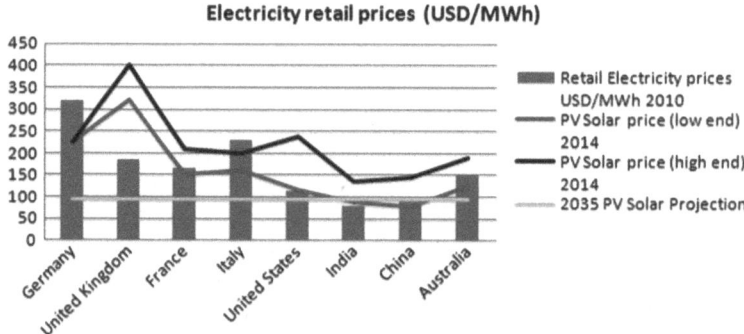

Fig. 5.17 Rooftop solar competitiveness (© OECD/IEA, Electricity 2012; WEC 2013)

countries. Within the next few years, the additional competitiveness of solar systems will create an opportunity in almost all countries of the world (Fig. 5.17).

The massive deployment of rooftop solar systems will therefore become a sound strategy. The traditional paradigm of centralized power plants with massive power output generating on an extended network infrastructure will thus be questioned.

5.3.3.4 A Solar Future?

Solar deployment will modify (it actually does already!) the way electrical systems operate, and consequently the way the electricity industry is organized. Economically, this is only a matter of time as the solar industry makes fantastic progress on a yearly basis. Future massive deployment will however bring with it a number of issues, both economic and technical.

First, utility-scale systems, as they grow more competitive, will naturally tend to substitute conventional plants, in particular coal power plants. Carbon taxes, when implemented, will accelerate this trend. The wider adoption of rooftop solar systems will also contribute to changing the power system economics by reducing the size of the wholesale market by a similar volume of the production capacity installed. A direct consequence will be the reduction of the market size for conventional generation, hence the restructuring of the sector.

Now, conventional power plants will continue to be required (at least on the short-term) because photovoltaic solar is by nature intermittent, and its output difficult to predict. As supply needs to equal demand at all times, conventional power plants, which only can be regulated, will thus continue to be needed, in particular for "peak load" and system balancing. This also means they will operate less continuously than before; their "load factor" will be reduced. This will affect prices because as long as these units are required, their economic balance must be assured. This means that wholesale prices will need to cover their fixed and variable costs, and provide an adequate level of margin for the operator. A first trend could thus be towards an increased volatility of prices, to better reflect the current cost of producing electricity at different times of the day, when there is an excess or a lack of renewable power on the grid. Then, capacity markets are also likely to develop, to generate fixed revenues to these conventional units in order to secure them on the grid.

Similarly, the higher penetration of intermittent photovoltaic solar will lead to an increase of transmission and distribution systems' operation costs, which are today borne by the final consumer thru retail electricity prices.

The operator will be forced to increase its "reserve" power in order to cope with the unpredictable and variable renewable production output. A number of innovations and measures should help smooth this effect. Weather forecasting will help increase the predictability of the solar contribution to the overall electricity demand. Grid codes will help enforce a higher level of regulation (voltage and frequency) at the borders of the solar farm. Now, costs should keep rising anyhow.

Then, conventional generation has traditionally been a key contributor to grid stability. With the restructuring of the conventional generation market, less of it will be available at a given point in time on the grid. Consequently, grid stability could be put at risk. This will lead to an extension of grid interconnections (at transmission level) across countries. The objective of those interconnections is indeed to bundle all production capacities together to ensure a higher level of reliability of the grid. These investments will have to be financed thru an increase of retail electricity prices.

The electrical network will also have to be managed differently. Traditionally, it has been designed to distribute energy flow in one direction (particularly at the distribution level). The network was built to accommodate the highest level of constraints, and consumption patterns were predictable. With the emergence of photovoltaic solar, the energy will now flow in all directions, since this type of generation source is generally connected on the distribution system, and not anymore on the transmission system like it used to be the case for conventional generation. This can lead to situations where the existing network is either durably oversized or congested by too much energy flow. Now, the return on investment of network extensions and upgrades could end up being highly questioned. Distribution utilities, operating at the lower end of the network and supplying electricity to small locations throughout the countryside, could indeed face situations where network costs of some regions become economically unbearable. In addition, the network operators' calculations on failure modes will become more complicated. Renewable energy flows will need to be integrated in the network model. This will lead to a massive increase of the possible failure modes on the network. Automation of the network will thus become paramount, as well as real-time reconfiguration, simulation and modeling solutions. These investments will have to be integrated into the overall system operation costs as well.

Finally, the deployment of rooftop solar systems will reduce the volume of electricity transiting thru the lines, hence the volume of electricity traded on which network and system operation costs are amortized. This will thus lead to a proportional increase of those costs for the final consumer, and this shall contribute to accelerate a further transition towards more competitive solutions such as rooftop solutions, since by nature they do not require any connection to the grid.

In the end, the higher penetration of solar (and wind) in the grid will thus lead to a number of issues, including a restructuring of the current power market, and most likely more variable prices with higher system operation costs included. Since the main issue is the lack of flexibility of solar energy, energy storage is the key element which, provided it would be economically sound to deploy it on a large scale, would help regulate solar contribution to the grid, and therefore reduce if not solve most of the adverse effects. The idea is to store all of what renewable sources can produce and regulate the output outside of the energy storage system according to demand requirements. While technical solutions do exist, they are not yet industrialized at the level to which they could be widely deployed. To date, only pumping stations possess large power storage capacities. These stations use electricity to pump water back up in a dam when there is little electrical consumption on the network, thus recharging the production capacity of the hydroelectric dam, which then can be used when consumption reaches its peak later in the day. Other storage technologies are also being developed. Energy storage can be done using compressed air systems, flywheels, electrochemical batteries (sodium-sulfur, lithium) and, potentially, hydrogen and semiconductors (© OECD/IEA, Storage 2014). All these technologies have different characteristics. A variety of constraints indeed apply to energy storage technologies, such as its overall power capacity, its discharge time duration, or its capacity to quickly react to a sudden increase in demand. The constraints are also very different between a utility-scale system and a home rooftop system. In the end, the advent of one or several industrialized energy storage technology/technologies will help cancel out the intermittence effect of solar energy and provide the necessary flexibility needed by the grid. Whether or not storage solutions will be able to replace completely conventional units for grid stabilization and system balancing remains a question. If they can, the complete transition of electricity production towards renewable energies, and in particular solar energy, would then be feasible.

Photovoltaic solar is thus a technology in development which offers massive opportunities. To date, 135 GW of production capacity has been installed worldwide. According to the International Energy Agency (2014), this volume should reach 1720 GW by 2030 and over 4670 GW by 2050. It would by then reach over 6000 TWh of electricity production, or 16% of global electricity production. This forecast is conservative, considering the massive trend towards solar energy. A more aggressive forecast from Greenpeace (2015) estimates that up to

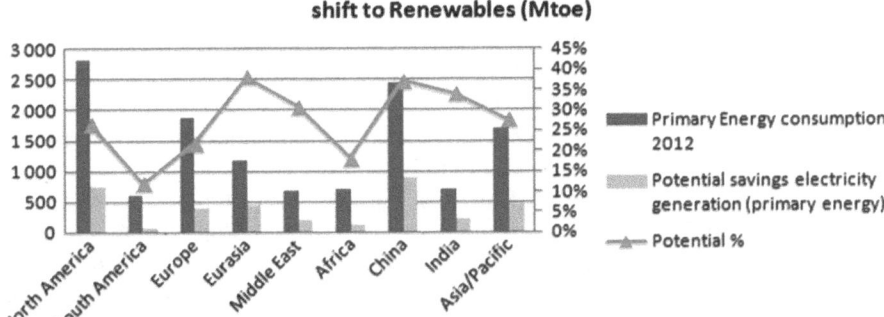

Fig. 5.18 Shift to renewable saving potential (© OECD/IEA, Explore 2014; © OECD/IEA, Solar 2014)

2800 GW could be installed by 2030, and 6700 GW by 2050. Economically interesting, decisive to other power technologies and environmentally friendly, solar electricity is a disruption to the electricity market which strongly challenges how the industry has operated for the last 70 years.

5.3.3.5 Regional Perspectives

With such a change of paradigm, the power and heat generation segment would undergo drastic changes and traditional utilities would be deeply transformed. The massive deployment of renewable energies (particularly solar) would solve the major energy waste issue in conventional power generation. The primary fossil energy consumed in 2010 for electricity production was 3600 Mtoe. The theoretical potential for primary energy saving here is thus 3600 Mtoe, through replacement of fossil fuels by renewable energies. This represents 29% of total primary energy demand.

Each geography's share of the potential engendered by renewable energies depends on the share of fossil fuels in the overall electricity generation mix of each country in that geography. Asia and North America have the highest stakes in terms of energy saving potential (Fig. 5.18).

5.4   Towards a Cleaner World: Fuel Switching Strategies

The rise of renewable energies will lead to a massive diminishment in the consumption of fossil fuels. The considerable waste from the very inefficient electricity generation process will come to an end. The switch to cost-effective (if not free) renewable-based electricity generation will require major fuel switching strategies.

5.4.1 More Efficient (Smarter) Cities

By 2050, 70% of the world population will live in cities. They will consume energy in their homes and in their workplaces, and they will mostly travel within cities. Cities of the twenty-first century are therefore the key physical places where energy will be consumed. In such cities, the spectacular consumption of energy for transportation and buildings will require to be optimized.

The concept of "smart cities" (European Parliament 2014; Fabrique de la cite 2014) relies heavily on optimizing the city and its transportation infrastructure. Today, 80% of transportation corresponds to short-distance travels, mostly within cities. Such travels account for around 40% of the total energy consumption of the sector (© OECD/IEA, Transport 2009). Most smart mobility programs thus aim to reduce travel's energy footprint within a city. These programs optimize the transportation system, with more collective transportation that reduces the energy intensity of individual transportation. They also redesign cities to reduce the need to use motorized transportation, and leverage ICTs to accelerate remote work and enable online shopping. Finally, they focus on substituting all conventional transportation with electric or carbon-free transportation. Future massive deployment of renewable energies in electricity production would help accelerate this substitution. In the end, the theoretical substitution of all short-distance travels by electric or carbon-free transportation could save up to 800 Mtoe of fossil fuels (© OECD/IEA, Transport 2009; © OECD/IEA, WEO 2012). For electric cars alone, current forecasts show a steep development. The number of electric cars could indeed reach six millions by 2020 (FS-UNEP 2016), and up to 25% of the worldwide fleet by 2040 (BNEF/Electric Vehicles 2016).

The next stage of more efficient cities regards buildings and their massive energy consumption. Of the total energy consumed in buildings, 40% comes from fossil fuels (primarily natural gas and coal). The pervasive deployment of renewable energies, such as solar energy (for heating) or geothermal energy, can help substitute these fuels; solutions already exist. Again, "smart cities" will be the ones which tackle these issues, favoring the development of distributed energy and smart grid, hence the substitution of fossil fuels consumption to electricity in buildings, as well as valuing the biodiversity potential of cities through parks, gardens, etc. The overall potential of energy that could be substituted in buildings amounts to 1100 Mtoe of primary energy (© OECD/IEA, WEO 2012).

The implementation of these concepts will naturally vary from one city to another. Some of the modern cities already offer a glimpse of what will be the cities of tomorrow. The massive use of solar energy, the deployment of cogeneration, connected transportation, electrical vehicles and even self-service bicycles are many examples of the development of cities designed with the consistent objective of becoming sustainable. The combination of these elements leads to a "smart city". Out of the 468 cities with more than 100,000 inhabitants in the European Union, 240 cities were already identified as "smart" by the European Parliament (2014).

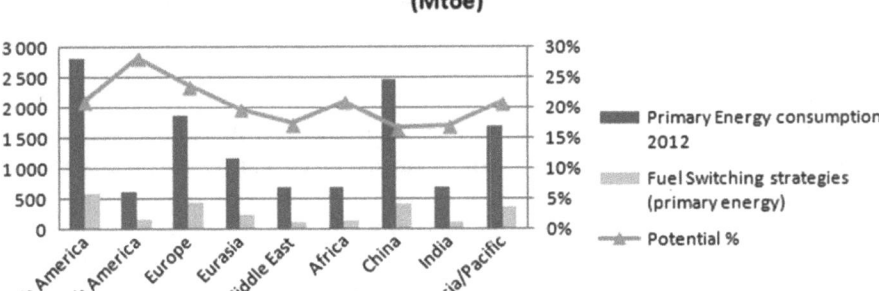

Fig. 5.19 Fuel switching strategies (© OECD/IEA, Technology Industry 2009; © OECD/IEA, Transport 2009; © OECD/IEA, WEO 2012)

5.4.2   A Greener Industry Sector

The industry sector is also sensitive to energy substitution as heating requires considerable amounts of energy, which can partially be supplied by renewable energies, essentially biomass. The International Energy Agency (2009) estimates that over 900 Mtoe of biomass energy could be used in the industry sector by 2050 in a high-base scenario (© OECD/IEA, Technology Industry 2009). This would represent an additional substitution of 700 Mtoe of fossil fuels compared to today's situation.

5.4.3   Regional Perspectives

Eventually, the deployment of renewable energies could lead to fuel switching strategies which would help save up to 2600 Mtoe. This means that up to 20% of the total primary energy demand (based on fossil fuels) could be saved by substituting conventional fuels (used for heating and transportation) with renewable energies.

The primary regions with high transportation and buildings intensity are North America, South America and Europe. Other regions such as China and the rest of Asia Pacific also present strong potential for moving away from conventional fuels in the transportation and buildings sectors, even though the share of the industry sector is stronger in their overall energy mix (Fig. 5.19).

5.5   The Energy Equation Can Be Solved

The world has a difficult energy equation to solve. On one hand, the development of new economies will lead to an increase of 35% in the energy demand in the next two decades. On another, CO_2 emissions need to be cut by 35% in the same time period in order to limit CO_2 concentration in the atmosphere to 450 ppm. A complete turnaround is required to solve the energy equation. Mature economies need to transition to greener and more efficient energy-consuming profiles, while new economies would grow their share of energy in the overall mix in an environmental friendly way.

This change is possible.

First, the potential for end use energy saving is massive. However, it requires the deployment of new technologies and changes to current energy consumption patterns. Up to 21% of the total primary energy, or 2700 Mtoe, could be saved.

Second, the volume of waste that our way of life generates is spectacular. Electricity generation in particular wastes a massive amount of primary resources. The pervasive deployment of efficient renewable electricity production, as a substitute to conventional fuels, could lead to up to 3600 Mtoe of primary energy saved, or 29% of the total primary energy consumed.

Finally, the emergence of competitive renewable energy sources of electricity and heat could lead to widespread fuel switching, with renewable energy substituting conventional fuels. The part of traditional transportation attributed to short-distance travel would be replaced by electricity-based transportation as a complement to extended smart mobility policies. Heating for buildings would be supplied by renewable energy sources. Also, a significant share of heat energy would be supplied by biomass in the industry sector. These substitutions alone could save up to 2600 Mtoe of primary energy, or a fifth of the total primary energy consumed.

Overall, the theoretical potential of primary energy savings is 8900 Mtoe. Of course, fuel switching strategies partially overlap with the energy efficiency measures which would reduce the demand on fossil fuels in end-use sectors. Nevertheless, the potential for reducing fossil fuels' energy consumption is considerable.

The graph below is a representation of energy consumption in 2012 and its forecast evolution in the "New Policy" scenario (© OECD/IEA, WEO 2012). It also shows the overall savings that could be made from improved end-user efficiency, using renewable energies for electricity production and fuel switching.

Clearly the main regions to focus on are:

- North America and Europe: these two regions present tremendous potential of energy savings,
- China and Asia Pacific: numerous energy efficiency opportunities already exist in these geographies (in particular, for electricity generation in China) and energy use there is expected to increase drastically in the next 20 years,
- India: energy consumption is projected to double in the coming two decades. While the savings potential in the country is low in absolute value, it could actually almost offset completely the energy increase (Fig. 5.20).

The energy equation can be solved provided energy consumption drops by 2400 Mtoe in the next 20 years, assuming the International Energy Agency's

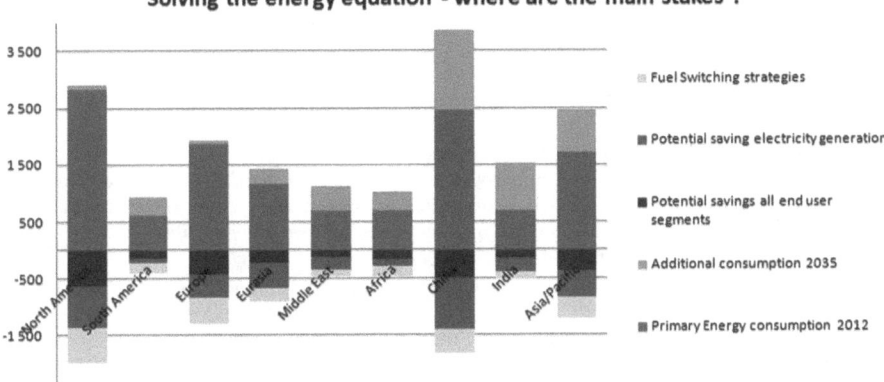

Fig. 5.20 Solving the energy equation (© OECD/IEA, Buildings 2013; © OECD/IEA, Efficient Buildings 2013; © OECD/IEA, Energy Efficiency 2013; © OECD/IEA, Explore 2014; © OECD/IEA, Motors 2011; © OECD/IEA, Solar 2014; © OECD/IEA, Statistics 2015; © OECD/IEA, Technology Industry 2009; © OECD/IEA, Transport 2009; © OECD/IEA, WEO 2012)

"New Policy" scenario, in which primary energy demand worldwide grows 35% from a 2010 base of 13,000 Mtoe. This corresponds to a 27% realization of the overall potential.

Reducing end-use consumption is a major requirement. A large share of the potential savings from improved end-use efficiency would however have to be realized to reach the above savings target. Considering the inertia of all sectors, this seems out of reach. The substitution of conventional electricity production with renewable energy and, more generally, the substitution of other conventional fuels used for heating or transportation with renewable energy are thus mandatory to reach the desired level of energy savings.

For the world to achieve 27% of the ultimate potential for savings would constitute a paradigm change. The change is not out of reach. Building renovation and active controls, renewables integration and waste heat management in industrial processes, smarter transportation, new motorization technologies, and electricity production transition to renewables are all actions that have already been launched in most countries where the stakes are high. Now, these actions need to be accelerated in order to realize the objective by 2035. The clock is ticking.

References

Barré B, Mérenne-Schoumaker B (2011) Atlas des énergies mondiales. Editions Autrement, Paris

BNEF/Electric Vehicles (2016) Electric vehicles to be 35% of global new car sales by 2040. http://about.bnef.com/press-releases/electric-vehicles-to-be-35-of-global-new-car-sales-by-2040/

Campbell R (2013) https://www.fas.org/sgp/crs/misc/R43343.pdf

CEA (2014) http://www.cea.fr/jeunes/themes/l-energie-nucleaire/l-essentiel-sur-la-fusion-nucleaire

CTA (2014) Center for transportation analysis. http://cta.ornl.gov/data/tedb33/Edition33_Full_
    Doc.pdf
Durand B (2007) Energie & Environnement, Les risques et les enjeux d'une crise annoncée.
    Collection Grenoble Sciences, EDP, Les Ulis
ESA (2014) http://esa.un.org/unpd/wup/Highlights/WUP2014-Highlights.pdf
European Parliament (2014) http://www.europarl.europa.eu/RegData/etudes/etudes/join/2014/
    507480/IPOL-ITRE_ET(2014)507480_EN.pdf
Fabrique de la cite (2014) http://www.lafabriquedelacite.com/fabrique-de-la-cite/site/fr/
    publications/pages/les_strategies_d_optimisation_urbaine_dans_7_villes_mondiales.htm,
    http://www.lafabriquedelacite.com/fabrique-de-la-cite/site/fr/publications/pages/etude_quel_
    role_pour_les_villes_dans_la_transition_energetique.htm
Favennec J-P, Mathieu Y (2014) Atlas mondial des energies. In: IPF Energies Nouvelles. Armand
    Collin, Paris
FEMP (2011) Federal Energy Management Program, US Department of Energy. http://www1.
    eere.energy.gov/femp/pdfs/eedatacenterbestpractices.pdf
Fraunhofer (2013) Levelized cost of electricity. https://www.ise.fraunhofer.de/en/publications/
    veroeffentlichungen-pdf-dateienen/studien-und-konzeptpapiere/study-levelized-cost-of-elec
    tricity-renewable-energies.pdf
FS-UNEP (2016) Frankfurt School UNEP collaborating center on climate and sustainable energy
    finance.   http://fs-unep-centre.org/sites/default/files/publications/
    globaltrendsinrenewableenergyinvestment2016lowres_0.pdf
Furfari S (2007) Le Monde et l'Energie, Enjeux géopolitiques. Editions Technip, Paris
GIEC/IPCC (2007) Rapport de synthèse. http://www.ipcc.ch/pdf/assessment-report/ar4/syr/ar4_
    syr_fr.pdf
Greenpeace (2015) Energy revolution. http://www.greenpeace.org/international/Global/interna
    tional/publications/climate/2015/Energy-Revolution-2015-Full.pdf
Hansen JP, Percebois J (2015) Energie: Economie et Politiques. Bruxelles, Ouvertures
    Economiques, De Boek
© OECD/IEA (2014) IEA Publishing. License: www.iea.org/t&c. As modified by V.Petit. http://
    www.iea.org/
© OECD/IEA, Buildings (2013) Transition to sustainable buildings. IEA Publishing. License:
    www.iea.org/t&c. As modified by V.Petit. http://www.iea.org/publications/freepublications/
    publication/transition-to-sustainable-buildings.html
© OECD/IEA, Coal (2014) CCS retrofit. IEA Publishing. License: www.iea.org/t&c. As modified
    by V.Petit. http://www.iea.org/publications/freepublications/publication/CCS_Retrofit.pdf
© OECD/IEA, Efficient Buildings (2013) Technology roadmap energy efficient building
    envelopes. IEA Publishing. License: www.iea.org/t&c. As modified by V.Petit. http://www.
    iea.org/publications/freepublications/publication/
    TechnologyRoadmapEnergyEfficientBuildingEnvelopes.pdf
© OECD/IEA, Electricity (2012) Electricity information. IEA Publishing. License: www.iea.org/
    t&c. As modified by V.Petit. http://www.iea.org/media/training/presentations/statisticsmarch/
    electricityinformation.pdf
© OECD/IEA, Energy Efficiency (2013) Energy efficient market report. IEA Publishing. License:
    www.iea.org/t&c. As modified by V.Petit. http://www.iea.org/publications/freepublications/
    publication/energy-efficiency-market-report-2013.html
© OECD/IEA, ETP (2012) Energy Technology Perspectives. IEA Publishing. License: www.iea.
    org/t&c. As modified by V.Petit. http://www.iea.org/publications/freepublications/publication/
    ETP2012_free.pdf
© OECD/IEA, Explore (2014) Explore. IEA Publishing. License: www.iea.org/t&c. As modified
    by V.Petit. http://www.iea.org/etp/explore/
© OECD/IEA, Motors (2011) Energy efficiency for electric systems. IEA Publishing. License:
    www.iea.org/t&c. As modified by V.Petit. https://www.iea.org/publications/freepublications/
    publication/EE_for_ElectricSystems.pdf

© OECD/IEA, Non-OECD (2012) Balance of non OECD countries. IEA Publishing. License: www.iea.org/t&c. As modified by V.Petit. http://www.iea.org/media/training/presentations/statisticsmarch/balancesofnonoecdcountries.pdf

© OECD/IEA, OECD (2012) Balance of OECD countries. IEA Publishing. License: www.iea.org/t&c. As modified by V.Petit. http://www.iea.org/publications/freepublications/

© OECD/IEA, Solar (2014) Technology roadmap solar photovoltaic energy. IEA Publishing. License: www.iea.org/t&c. As modified by V.Petit. https://www.iea.org/media/freepublications/technology roadmaps/solar/TechnologyRoadmapSolarPhotovoltaicEnergy_2014edition.pdf

© OECD/IEA, Statistics (2015) Statistics. IEA Publishing. License: www.iea.org/t&c. As modified by V.Petit. http://www.iea.org/Sankey/#?c=World&s=Final consumption

© OECD/IEA, Storage (2014) Technology roadmap energy storage. IEA Publishing. License: www.iea.org/t&c. As modified by V. Petit. http://www.iea.org/publications/freepublications/publication/technology-roadmap-energy-storage-.html

© OECD/IEA, Technology Industry (2009) Technology Industry. IEA Publishing. License: www.iea.org/t&c. As modified by V. Petit. http://www.iea.org/publications/freepublications/publication/industry2009.pdf

© OECD/IEA, Transport (2009) Transport. IEA Publishing. License: www.iea.org/t&c. As modified by V. Petit. http://www.iea.org/publications/freepublications/publication/transport2009.pdf

© OECD/IEA, WEO (2012) World energy outlook. IEA Publishing. License: www.iea.org/t&c. As modified by V. Petit. http://www.worldenergyoutlook.org/publications/weo-2012/

IPCC (2007) Intergovernmental Panel on Climate Change. https://www.ipcc.ch/publications_and_data/ar4/wg3/en/ch5.html

Irena (2012) International Renewable Energy Agency. http://www.irena.org/DocumentDownloads/Publications/RE_Technologies_Cost_Analysis-SOLAR_PV.pdf

JTLU (2010) Journal of Transport and Land Use. http://transportsdufutur.typepad.fr/files/mobility_integrated.pdf

L'Expansion (2012) http://energie.lexpansion.com/energie-nucleaire/la-cogeneration-nucleaire-une-formidable-economie-d-energie_a-32-7217.html

Lazard (2014) Levelized costs of energy. http://www.lazard.com/media/1777/levelized_cost_of_energy_-_version_80.pdf

Manicore (2014) http://www.manicore.com/anglais/documentation_a/solar.html

National Academies (2009) http://www.nationalacademies.org/includes/G8+5energy-climate09.pdf

OMM (2014) Organisation Météorologique Mondiale. http://www.wmo.int/pages/mediacentre/press_releases/pr_1002_fr.html

Parmentier B (2009) Nourrir l'humanité: Les grands problèmes de l'agriculture mondiale au xxie siècle. Ed. La Découverte, Paris

Passivhaus (2014) http://www.passivhaus.fr/10.html

Road Transport (2012) India road transport year book 2011–2012. http://morth.nic.in/showfile.asp?lid=1131

Schneider Electric (2014) http://www.schneider-electric.com/solutions/ww/en/seg/4663977-buildings/4872918-real-estate-office-buildings

Suez (2014) http://www.gdfsuez.com/engagements/recherche-et-technologies/captage-stockage-du-co2/

UK DoT (2009) UK Department of Transportation. http://webarchive.nationalarchives.gov.uk/20110109134413/http://www.dft.gov.uk/adobepdf/162469/221412/190425/220778/trends2009.pdf And https://www.gov.uk/government/uploads/system/uploads/attachment_data/file/342160/nts2013-01.pdf

UN (2014) United Nations. http://www.un.org/en/development/desa/news/population/world-urbanization-prospects-2014.html

US DoT (2009) US Department of Transportation. http://nhts.ornl.gov/2009/pub/stt.pdf
Vendryes (2001) http://ecolo.org/documents/documents_in_french/fusion-vendryes-01.htm
WEC (2013) http://www.worldenergy.org/publications/2013/world-energy-perspectivecost-of-energy-technologies/

Towards 2100

6

6.1 Historical Continuities Shape the World of Tomorrow

Considerable historical continuities have shaken the balance established previously. The world we live in today has already undergone several major transformations. As always, these changes are difficult to perceive as they take several decades to fully realize. When immersed in the daily actions of human life, it is often difficult to take some distance and look at the historical changes happening before one's eyes. These changes however will what will be remembered from our time, and it will become obvious, when looking at this period from far in the future, that these continuities have shaped the new world we are entering.

The speed at which these changes occur has increased considerably. This is a new reality. In less than a century, in the twentieth and midway into the twenty-first century, the world's population will have increased from 2.5 billion people to more than nine billion. Global GDP will have been multiplied by 40, completely disrupting what humanity has experienced for centuries. The blooming of the world population and its living standards are due to industrial revolutions and to the fantastic development of health services, which pulled mankind out of a subsistence economy.

This prodigious development created considerable needs in terms of energy. The world that shaped up after the Second World War based its progress upon the massive use of fossil fuels. Oil and natural gas from the Middle East or from Russia complemented coal, which had been the main source of energy that powered the first industrial revolutions in Europe and North America during the nineteenth century. Superpowers such as the United States or the United Socialist Soviet Republics ensured their control over these resources and their procurement in order to maintain their status and the domination they exercised over the rest of the world.

Energy consumption is increasing today at breakneck speed, but paradoxically there have never been so many resources available. Actually, unconventional fossil

© Springer International Publishing AG 2017   169
V. Petit, *The Energy Transition*, DOI 10.1007/978-3-319-50292-2_6

fuels changed completely the situation. The American continent is now independent from an energy standpoint. China, which is on the verge of exhausting its coal
resources, could find in shale gas an alternative to ensure the security of its energy
supply. Possible historical changes could occur: evolutions in the United States
foreign policy with regards to the Middle East, new partnerships between countries
surrounding the Arabian Sea, or on the eastern borders of Russia are all new
possible historical pathways.

The colossal growth of energy consumption has also become a threat to global
climate sustainability. The world's economic development is based on fossil fuel
consumption. Our energy needs are already considerable, and keep increasing.
Already, drastic climate changes seem to be irremediable. These historical continuities will weigh heavily in the twenty-first century.

Mitigating these changes will require in the coming twenty years a reduction of
greenhouse gas emissions by 35%. On the flip side, energy demand is set to increase
35% over the same time period. The world is thus facing an impossible equation.
Our generation needs to solve this equation in order to prepare a sustainable future
for the generations that follow.

Thankfully, it is possible to reduce our energy footprint. First of all, the potential
for improving energy efficiency worldwide is considerable. Around 21% of primary
energy could be saved by the effective deployment of energy efficiency measures.
Here, the issue is not so much discovering new technologies as deploying efficiently
existing ones. The gaps between what can best be done and what generally exists is
colossal in all sectors. Upgrading each sector (Industry, Buildings, Transportation)
to what can best be achieved in terms of energy consumption would lead to
significant savings. One main hurdle to such achievement is inertia. The multitude
of players, each with its own set of interests it wishes to defend; the divergence and
sometimes opposition of nations' interests; the unwillingness to replace existing
energy-inefficient assets. All these lead to a global lethargy that could delay the
realization of a smaller collective energy footprint.

Renewable energies, in particular photovoltaic solar, present a historic opportunity to deeply modify the energy mix for electricity production. Associated with
energy storage, renewable energies could substitute fossil fuels to a very large
extent in the coming years; the substitution rate (not counting hydro power) is less
than 10% today. Using renewable energies for electricity production could also help
solve the massive issue of waste from electricity generation. The potential adds up
to around 29% of total primary energy consumption. Finally, the deployment of
renewable energies, in particular rooftop solar systems, could push the various end-
use sectors to massively switch to renewable energy. This would represent an
additional potential of 20% of primary energy savings. In the end, up to 70% of
total primary energy could either be saved or be produced with renewable energies;
only 19% needs to be saved to fully realize the 450 scenario, which aims to limit the
CO_2 concentration in the atmosphere to 450 ppm. The world's energy equation can
thus be cracked, and a renewable and sustainable future is possible.

One must believe in the ability of humankind to control its fate and overcome the
challenges it creates for itself. This book had no other intention but to shed light on

these challenges, and list out the variety of existing solutions that could be deployed.

6.2   What Happens Next?

By the end of the century, the world population will have reached its peak of above nine billion people. Then, the world population will stop increasing, and it will even start to decline in a number of regions. One of the main sources of economic growth will be exhausted.

The historical economic catch-up of new economies will be achieved in most places. It is already happening in Asia. It should start later in Africa, and the continent will become the last oasis of growth of the planet. One can imagine that all these transitions will be over by the end of this century. Beyond all vicissitudes which occur along this transition, the picture of a global interconnected village is already shaping up in people's minds (Rifkin 2012). Homogenization of cultures and economic needs will precede that of living standards. There is no doubt on the outcome.

The world will also be older. The share of those 65 and above will tend towards one third of the world population, with strong differences across regions. This will encourage a further blending of populations across regions, as young people will emigrate more frequently towards regions in great need of workforce. This evolution will shake if not dismantle the ancient principles of solidarity, both within a nation and across generations, as well as the foundations on which the concept of nations is built.

Finally, technology will help deploy massively renewable energies to create an all-electric world, one where energy will be free.

This fantastic evolution is just ahead of our time! By the end of the century, the world will be without economic growth, saturated of wealth, global, interconnected, and older than today. These evolutions are not science fiction; they represent historical continuities which can already be identified.

One question that will arise is: what will be the meaning of life in such a world? What will drive such a population, one that is urban, wealthy, interconnected, blended, older, and living in greater harmony with nature?

And what will happen to the capitalist system on which our global economic system is based? Capitalism brought benefits to the world population. One facet of it is the accumulating of wealth. When global wealth will not increase anymore, the share of what will be left will be under pressure. Economic rents will end up being shared within small groups. How will the vast majority of the world population, which will have no access to it, act in a world without perspectives? Already, many critics are questioning the balance of economic power between rich and poor and the way international institutions such as the International Monetary Fund operate (Stiglitz 2002).

What will happen to the hope to grow, the true engine of humankind in the past few centuries? What can an ageing population, freed from the constraints that nature imposed for centuries, hope for? Will economic stasis be the ultimate state of development? For centuries, despite being chained to a subsistence economy, people all over the world have hoped for something better. From time to time, an event changed their life, for better or for worse. The industrial revolutions suddenly propelled a continuous rise in their living standards over two centuries. Democracies shaped up, and people started to believe that they had been elected to rule the planet and that each of us were able to achieve anything, thanks to technology. In less than four generations, the world will be saturated and old. The picture of its evolution will be frozen. The hope for growth and elevation, as we knew it, will naturally fade away.

Can we live in such a world, or will humankind set itself a new challenge? What's next?

References

Rifkin J (2012) La troisième révolution industrielle. Editions les liens qui libèrent, Paris
Stiglitz J (2002) La grande désillusion. Fayard, Paris

Zeitfracht Medien GmbH
Ferdinand-Jühlke-Straße 7
99095 Erfurt, Deutschland
produktsicherheit@kolibri360.de